本书为以下项目的部分成果：

南京大学外国语学院"双一流"学科建设项目

全国科学技术名词审定委员会重点项目"中国世界级非物质文化遗产术语英译及其译名规范化建设研究"

教育部学位中心 2022 年主题案例项目"术语识遗：基于术语多模态翻译的中国非物质文化遗产对外译介与国际传播"

南京大学 - 江苏省人民政府外事办公室对外话语创新研究基地项目

江苏省社科基金青年项目"江苏世界级非物质文化遗产术语翻译现状与优化策略研究"（19YYC008）

江苏省社科基金青年项目"江苏世界级非遗多模态双语术语库构建研究"（23YYC008）

南京大学暑期社会实践校级特别项目"讲好中国非遗故事"校园文化活动

中国世界级非遗文化悦读系列·寻语识遗
丛书主编 魏向清 刘润泽

中国传统桑蚕丝织技艺·宋锦
（汉英对照）

刘润泽 董晓娜 主编

Traditional Chinese Sericulture and Silk Craftsmanship: Song Brocade

南京大学出版社

参与人员名单

丛书主编 魏向清 刘润泽
主　　编 刘润泽 董晓娜
翻　　译 王朝政
译　　校 Zhujun Shu　Benjamin Zwolinski
学术顾问 沈芝娴
出版顾问 何宁　高方
中文审读专家（按姓氏拼音排序）
　　　　　　陈俐　丁芳芳　王笑施
英文审读专家 Colin Mackerras　Leong Liew
参编人员（按姓氏拼音首字母排序）
　　　　　　陈楠　陈逸凡　范文澜　冯雪红　高敏
　　　　　　江娜　蒋思佳　梁鹏程　刘争妍　秦曦
　　　　　　孙澳　孙文龙　吴小芳
手　　绘 陶李
实 物 图 苏州钱小萍宋锦大师工作室
知识图谱 王朝政
中国历史纪年简表 王朝政
特别鸣谢 江苏省非物质文化遗产保护研究所
　　　　　　苏州钱小萍宋锦大师工作室

编者前言

2019年秋天开启的这次"寻语识遗"之旅，我们师生同行，一路接力，终于抵达了第一个目的地。光阴荏苒，我们的初心、探索与坚持成为这5年奔忙的旅途中很特别，也很美好的回忆。回望这次旅程，所有的困难和克服困难的努力，如今都已经成为沿途最难忘的风景。

这期间，我们经历了前所未有的自主性文化传承的种种磨砺，创作与编译团队的坚韧与执着非同寻常。古人云，"唯其艰难，方显勇毅；唯其磨砺，始得玉成"。现在即将呈现给读者的是汉英双语对照版《中国世界级非遗文化悦读系列·寻语识遗》丛书（共10册）和中文版《中国世界级非遗文化悦读》（1册）。书中汇聚了江苏牵头申报的10项中国世界级非物质文化遗产项目内容，我们首次采用"术语"这一独特的认知线索，以对话体形式讲述中国非遗故事，更活泼生动地去诠释令我们无比自豪的中华非遗文化。

2003年，联合国教科文组织（UNESCO）第32届会议正式通过了《保护非物质文化遗产公约》（以下简称《公约》），人类非物质文化遗产保护与传承进入了全新的历史时期。20多年来，世界"文化多样性"和"人类创造力"得到前所未有的重视和保护。截至2023年12月，中国被列入《人类非物质文化遗产代表作名录》的项目数量位居世界之首（共43项），是名副其实的世界非遗大国。正如《公约》的主旨所述，非物质文化遗产是"文化多样性保护的熔炉，又是可持续发展的保证"，中国非遗文化的世界分享与国际传播将为人类文化多样性注入强大的精神动力和丰富的实践内容。事实上，我国自古就重视非物质文化遗产的保护与传承。"收百世之阙文，采千载之遗韵"，现今留存下来的卷帙浩繁的文化典籍便是记录和传承非物质文化遗产的重要载体。进入21世纪以来，中国政府以"昆曲"申遗为开端，拉开了非遗文化国际传播的大幕，中国非遗保护与传承进入国际化发展新阶段。各级政府部门、学界和业界等多方的积极努力得到了国际社会的高度认可，中国非遗文化正全面走向世界。然而，值得关注的是，虽然目前中国世界级非物质文化遗产的对外译介与国际传播实践非常活跃，但在译介理据与传播模式方面的创新意识有待加强，中国非遗文化的国际"传播力"仍有待进一步提升。

《中国世界级非遗文化悦读系列·寻语识遗》这套汉英双语丛书的编译就是我们为中国非遗文化走向世界所做的一次创新译介努力。该编译项目的缘起是南京大学翻译专业硕士教育中心特色课程"术语翻译"的教学实践与中国文化外译人才培养目标计划。我们秉持"以做促学"和"全过程培养"的教学理念，探索国别化高层次翻译专业人才培养的译者术语翻译能力提升模式，尝试走一条"教、学、研、产"相结合的翻译创新育人之路。从课堂的知识传授、学习，课后的合作研究，到翻译作品的最终产出，我们的教研创新探索结出了第一批果实。

汉英双语对照版丛书《中国世界级非遗文化悦读系列·寻语识遗》被列入江苏省"十四五"时期重点图书出版规划项目，这是对我们编译工作的莫大鼓励和鞭策。与此同时，我们受到来自国际中文教育领域多位专家顾问的启发与鼓励，又将丛书10册书的中文内容合并编成了一个合集《中国世界级非遗文化悦读》，旨在面向国际中文教育的广大师生。2023年夏天，我们这本合集的内容经教育部中外语言交流合作中心教研项目课堂试用，得到了非常积极的反馈。这使我们对将《中国世界级非遗文化悦读》用作非遗文化教材增添了信心。当然，这个中文合集版本也同样适用于国内青少年的非遗文化普及，能让他们在"悦读"过程中感受非遗文化的独特魅力。

汉英双语对照版丛书的编译理念是通过"术语"这一独特认知路径，以对话体形式编写术语故事脚本，带领读者去开启一个个"寻语识遗"的旅程。在每一段旅程中，读者可跟随故事里的主人公，循着非遗知识体系中核心术语的认知线索，去发现、去感受、去学习非遗的基本知识。这样的方式，既保留了非遗的本"真"知识，也彰显了非遗的至"善"取向，更能体现非遗的大"美"有形，是有助于深度理解中国非遗文化的一条新路。为了让读者更好地领会非遗知识之"真善美"，我们将通过二维码链接到"术语与翻译跨学科研究"公众号,计划陆续为所有的故事脚本提供汉语和英语朗读的音频,并附上由翻译硕士专业同学原创的英文短视频内容,逐步完成该丛书配套的多模态翻译传播内容。这其中更值得一提的是,我们已经为这套书配上了师生原创手绘的核心术语插图。这些非常独特的用心制作融入了当代中国青年对于中华优秀传统文化的理解与热爱。这些多模态呈现的内容与活泼的文字一起将术语承载的厚重知识内涵,以更加生动有趣的方式展现在读者面前,以更加"可爱"的方式讲好中国非遗故事。

早在10多年前,全国高校就响应北京大学发起的"非遗进校园"倡议,成立了各类非遗文化社团,并开展了很多有益的活动,初步提升了高校学生非遗文化学习的自觉意识。然而,我们发现,高校学生群体的非遗文化普及活动往往缺乏应有的知识深度,多限于一些浅层的体验性认知,远未达到文化自知的更高要求。我们所做的一项有关端午非遗文化的高校学生群体调研发现,大部分高校学生对于端午民俗的了解较为粗浅,相关非遗知识很是缺乏。试问,如果中国非遗文化不能"传下去",又怎能"走出去"?而且,从根本上来说,没有对自身文化的充分认知,是谈不上文化自信的。"求木之长者,必固其根本;欲流之远者,必浚其泉源。"中国世界级非遗文化的对外译介与国际传播要解决的关键问题是培养国人尤其是青少年的非遗文化自知,形成真正意义上基于文化自知的文化自信,然后才有条件由内而外，加强非遗文化的对外译介与国际传播。非遗文化小书的创新编译过程正是南京大学"非遗进课堂"实践创新的成果，也是南大翻译学子学以致用、培养文化自信的过程。相信他们与老师一起探索与发现，创新与传承，译介与传播的"寻语识遗"之旅定会成为他们求学过程中一个重要的精神印记。

我们要感谢为这10个非遗项目提供专业支持的非遗研究与实践方面的专家，他们不仅给我们专业知识方面的指导和把关，而且也深深影响和激励着我们，一步一个脚印，探索出一条中国非遗文化"走出去"和"走进去"的译介之路。事实上，这次非常特别的"寻语识遗"之旅，正是因为有了越来越多的同行者而变得更加充满希望。最后，还要特别感谢南京大学外国语学院给了我们重要的出版支持，特别感谢所有参与其中的青年才俊，是他们的创意和智慧赋予了"寻语识遗"之旅始终向前的不竭动力。非遗文化悦读系列是一个开放的非遗译介实践成果系列，愿我们所开辟的这条"以译促知、以译传通"的中国非遗知识世界分享的实践之路上有越来越多的同路人，大家携手，一起为"全球文明倡议"的具体实施贡献更多的智慧与力量。

目 录
Contents

百字说明　　A Brief Introduction

内容提要　　Synopsis

知识图谱　　Key Terms

衣冠南渡　　Southward Migration of the Nobles ·················· 001

重锦　　Exquisite Brocade ·················· 014

细锦　　Fine Brocade ·················· 026

几何纹　　Geometric Design ·················· 033

天华锦　　*Tianhua* Brocade ·················· 042

灯笼锦　　Lantern Brocade ·················· 055

福娃纹　　Fuwa Design ·················· 065

海水江崖纹　　Sea-and-Mountain Pattern ·················· 073

脚踏缫丝车　　Treadle Reeling Machine ·················· 084

精练　　Silk Scouring ·················· 095

茜草　　Madder ·················· 105

靛青　　Indigo ·················· 114

攀华　　*Panhua* Process ·················· 123

活色 *Huose* Technique ·· 134

苏州织造署　Suzhou Weaving and Dyeing Bureau ················· 143

立桥　丝账房　Bridge Standby and Silk Trader ···················· 152

雅文化　Elegance Culture ·· 162

结束语　Summary ·· 173

中国历史纪年简表　A Brief Chronology of Chinese History ········ 175

百字说明

宋锦是中国三大名锦之一，其织造技艺是传统桑蚕丝织技艺的重要代表。宋锦主要产自苏州，故又称苏州宋锦。宋锦纹样古朴雅致，活色技术独特，具有典型的宋代织锦风格，被誉为"锦绣之冠"。宋锦织造技艺成于南宋，盛于明清。2009年包括杭罗、宋锦、缂丝等在内的中国传统桑蚕丝织技艺被列入联合国教科文组织的《人类非物质文化遗产代表作名录》。

A Brief Introduction

As one of the three famous brocades in China, Song brocade is highly representative by its exquisite craftsmanship. It is mainly produced in Suzhou, hence also known as Suzhou Song brocade. With its distinctive style of brocade weaving formed in the Song Dynasty, Song brocade is honoured as the "Crown of Brocade" in China. Its patterns feature simplicity and elegance animated by changing colours and hues, the richness of which is achieved through the unique *huose* technique. The weaving skills of Song brocade took shape in the Southern Song Dynasty and embraced further development and wider application in the Ming and Qing dynasties. In 2009, traditional sericulture and silk craftsmanship of China, Hangluo gauze, Song brocade and Kesi included, was inscribed on the Representative List of the Intangible Cultural Heritage of Humanity by the UNESCO.

内容提要

小龙的父亲龙教授是丝织研究专家。他参加宋锦研讨会后,给小龙带回了一件礼物——宋锦围巾。由此,父子俩展开了有关宋锦及其历史文化的对话。之后,小龙和大卫一起专程去苏州丝绸博物馆参观,学习宋锦织造技艺的基础知识。

Synopsis

Prof. Long, Xiaolong's father, is a silk-weaving researcher. He brought back a scarf made of Song brocade for Xiaolong after attending a seminar on Song brocade. Hence a discussion about the history and culture of Song brocade started. After that, Xiaolong and his friend David visited Suzhou Silk Museum and gained some basic knowledge of the weaving skills of Song brocade.

知识图谱
Key Terms

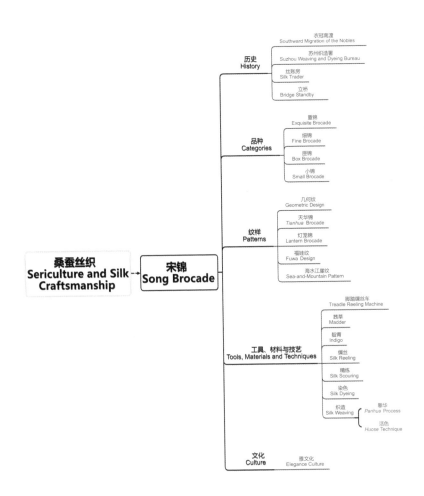

衣冠南渡

> 龙教授去参加一年一度的宋锦学术研讨会,给小龙带回一条宋锦围巾。

龙教授：小龙,这是宋锦围巾,爸爸带给你的礼物,看看喜不喜欢?

小　　龙：谢谢爸爸。太喜欢了,花色真好看。不过我不明白,宋锦明明是苏州产的织锦,为什么不叫苏锦,而叫宋锦呢?难道是跟宋朝有关系吗?

龙教授：问得好。宋锦产自苏州,和宋朝有渊源。

小　　龙：什么渊源?爸爸,快讲讲。

龙教授：这个要先从锦说起。

小　　龙：我听说锦在古代算奢侈品,一般只有皇室和贵族才能享用。

龙教授：没错。据史料记载,锦起源于3000多年前的周朝,……是一种多彩提花织物，也是丝织品中最复杂的一个品种。锦本身又分为经线显花和纬线显花两种不同的织法。早期的锦通常是经线显花，唐朝时出现了纬线显花的工艺。宋锦在此基础上进一步发展，有了经线和彩纬联合显花的新工艺。

小　　龙：这是织锦技术的重大进步吧？

龙教授：当然了。经过长期发展，织锦工艺在北宋时期得到了全面提高。到南宋时期，织锦又吸收了花鸟画的特点，大大增加了织锦的艺术性，形成了一种色彩复杂、清新雅致的新式风格。宋锦的特点是在几何形骨架设计上添加各种吉祥纹样，形成连续有序的图案。这种独特的织锦风格是在宋朝形成的，宋锦由此得名。宋锦当时主要用于宫廷服饰和书画装帧。

小　　龙：那南宋以前宫廷里主要用什么锦呢？

龙教授：在那之前，宫廷里大多用蜀锦。蜀锦历史最悠久，素有"天下母锦"的美称。

小　　龙：从名称来看，蜀锦的产地应该是在四川？

龙教授：对。事实上，织锦从蜀锦到宋锦的发展见证了古代织锦中心的迁移。而织锦中心的迁移与宋朝南迁的……

这段历史有关。

小　　龙：宋朝南迁的历史？我记得历史书上说，北宋灭亡后皇室南下，迁都到了临安，也就是今天的杭州。"南宋"这个叫法也是这么来的，但这跟织锦中心的迁移有什么关系呢？

龙教授：有关系啊。当时，人们为了逃避战乱，也跟着迁去江南地区，这就是历史上有名的第三次"衣冠南渡"。

小　　龙：衣冠南渡？"衣冠"不是衣服和帽子吗？

龙教授：这个不能简单地从字面上理解。在汉语中，"衣

衣冠南渡
Southward Migration of the Nobles

"冠"还可以指代有身份、有地位的人。中国历史上有三次衣冠南渡，极大地影响了中华文明的发展格局。

小　　龙：这三次衣冠南渡都发生在什么时期呢？

龙教授：第一次发生在公元316年西晋灭亡时，当时的统治政权从河南洛阳迁到长江以南的建康，就是今天的南京。这是第一次中原地区大规模的人口南迁，意味着中原文化开始向江南发展。

小　　龙：那第二次呢？

龙教授：第二次是在唐朝末年，也是因为战乱。这次人口南迁持续了很长时间，前后约有100万人从北方地区迁到南方。这次南迁从根本上改变了中国人口的布局，南方的人口规模第一次超过了北方地区。

小　　龙：爸爸，我可以这么理解吗？历史上的衣冠南渡，实际上是指大规模的人口南迁，它们客观上带动了江南的经济和文化发展。

龙教授：没错。第三次衣冠南渡后，随着社会秩序的恢复和稳定，江南成为新的经济和文化中心，许多江南名城就是在那时发展起来的。

小　　龙：苏州好像是南宋最繁华的城市之一。

龙教授：对。正是在那个时期，宋锦逐步形成了规模化生产，主要产地就在苏州。后来苏州慢慢就成了织锦中心。

小　龙：织造业在苏州复兴，与皇室对织锦的需求有很大关系吧？

龙教授：没错。宋朝迁都后，就专门在苏州设立了官方的织造机构，网罗全国各地的优秀织工，织造工艺精湛、风格雅致的新式织锦。

小　龙：难怪叫宋锦，宋朝的织锦最出名了。

龙教授：那倒不能这么说。事实上，宋锦出名是在明清时期，尤其是清朝。到了清代，苏州织锦工艺已经远近闻名，而且宋代古朴清雅的纹样备受青睐，逐渐就有了苏州宋锦的说法。现在苏州宋锦、四川蜀锦，还有南京云锦并称中国三大名锦。

小　龙：没想到宋锦还有这么悠久的历史。

Southward Migration of the Nobles

> Back from an annual seminar on Song brocade, Prof. Long brought a Song brocade scarf for Xiaolong.

Prof. Long: Hey, Xiaolong, look what I brought you. A scarf made of Song brocade from Suzhou. Do you like it?

Xiaolong: Thanks, Dad! The colours and patterns are so stunning. I love it. But I am a bit curious. It's a local speciality from Suzhou, right? Then why is it called Song brocade instead of Su brocade? Does it have something to do with the Song Dynasty?

Prof. Long: Sure. The name itself shows its historical relations with the Southern Song Dynasty.

Xiaolong: What kind of relations? I can't wait to hear more about it.

Prof. Long: Well, to start with, we should go back to the Chinese history of brocade first.

Xiaolong: Ah, I know something about it. It's said that in ancient China, brocade was considered a luxury only used by the imperial family and the nobles.

Prof. Long: That's true. According to historical records, brocade originated in the Zhou Dynasty over 3,000 years ago. It's a type of multicolour fabric and is considered the most intricate kind among silk textiles. Brocade can be further divided into two kinds based on its weaving methods, namely, the warp-faced brocade and the weft-faced brocade. In the early stages, brocade predominantly featured the warp-faced technique, and the weft-faced technique emerged later during the Tang Dynasty. As for Song brocade, it integrates creatively both of them.

Xiaolong: So, this marks a significant progress in brocade weaving, right?

Prof. Long: Exactly. By the time of the Northern Song Dynasty, brocade weaving in China had been practised continuously for a long time and the craftsmanship then had reached a new level. Later, in the Southern Song Dynasty, the technique of brocade weaving went a step further by assimilating the artistry of floral and bird paintings, with its artistic value greatly enhanced. It gave rise to a new style of brocade characterised by intricate colours and refreshing elegance. The brocade at that time was often found elaborately patterned by items with auspicious connotations. Those decorative items appear repeatedly and neatly within a geometric framework with nuances in design. Since this unique style of brocade appeared in the Song Dynasty, it came to be known as Song brocade. At that time, Song brocade were mainly for making imperial costumes and mounting calligraphy or painting works.

Xiaolong: I see. I have a question then. Before the Southern Song Dynasty, what kind of brocade was mainly used by the imperial family?

Prof. Long: Before that, Shu brocade was primarily used in the imperial court. It has the longest history and is often called the "Mother of Brocade".

Xiaolong: As its name could tell, Shu brocade was produced in Sichuan province, right?

Prof. Long: Correct! The development from Shu brocade to Song brocade actually witnessed the relocation of brocade weaving centre. This brings us to the history of the southward migration in the Song Dynasty.

Xiaolong: Oh, I know that. According to the history book, after the fall of the Northern Song Dynasty, the imperial family moved the capital to the south and settled down in Lin'an, the city of Hangzhou today, and hence the name "the Southern Song Dynasty". The migration was the prelude to the shift of the brocade weaving centre back then, I guess?

Prof. Long: You're right. At that time, many people fled to the south of the Yangtze River to escape from the wars and riots. That's the well-known historical event, the third southward migration of the nobles in history, also known in Chinese as the third *yiguan nandu*（衣冠南渡）.

Xiaolong: What? Everybody knows that *yi*（衣）and *guan*（冠）in Chinese refer to clothes and hats respectively. Then, how can clothes and hats move southwards?

Prof. Long: Well, that's only half of the story. Here, clothes and hats refer to the people of high status. In fact, there were three rounds of southward migrations of the nobles in Chinese history, which contributed significantly to the reshaping of the Chinese cultural geography.

Xiaolong: So, when did these events take place?

Prof. Long: The first southward migration occurred following the downfall of the Western Jin Dynasty when the ruling power moved from Luoyang, Henan province to Jiankang, in the south of the Yangtze River. Jiankang is today's Nanjing, in Jiangsu province. This marks the first large-scale migration from the Central Plains to the south. It also means the Chinese cultural centre began to move towards the south of the Yangtze River.

Xiaolong: I see. Then what about the second migration?

Prof. Long: The second migration happened at the end of the Tang Dynasty and lasted for a long time. About one million people moved southwards, which fundamentally changed the population distribution in China. As a result, the population in the south was as many as or even more than that of the north for the first time.

Xiaolong: These southward migrations in history were large-scale population shifts. They virtually promoted the economic and cultural development in the south of the Yangtze River. Am I right, Dad?

Prof. Long: Absolutely. After the third migration, the Southern Song Dynasty restored its social order and stability. The south of the Yangtze River replaced the Central Plains to become the new economic and cultural centre where many renowned cities were growing and developing quickly.

Xiaolong: Suzhou is one of them, right? I guess it became prosperous then.

Prof. Long: Yes, indeed! It was during the Southern Song Dynasty that Song brocade gradually came into large-scale production, with its main production centre located in Suzhou. Later on, Suzhou gradually became the centre of brocade weaving.

Xiaolong: I guess the revival of the textile industry in Suzhou was very much related to the growing imperial demand for brocade.

Prof. Long: That's true. In fact, after the Song Dynasty relocated its capital, the court established an official weaving bureau in Suzhou. The skilful weavers nationwide were gathered to produce a new type of brocade with exquisite craftsmanship and elegant style. That's the Song brocade we're talking about.

Xiaolong: So, it was during the Song Dynasty that Song brocade reached the height of its fame, right?

Prof. Long: Not exactly. Actually, it was during the Ming and Qing dynasties, particularly the Qing Dynasty, that

Song brocade gained widespread fame. In the Qing Dynasty, the brocade-weaving skills in Suzhou became well-known far and wide. The simple and elegant patterns designed in the Song Dynasty were highly favoured, giving rise to the term "Suzhou Song brocade". Now, Suzhou Song brocade, Sichuan Shu brocade and Nanjing Yunjin brocade are honoured as China's "three top brocades ".

Xiaolong: I never imagined that there would be such a long and rich history behind Song brocade.

重锦

> 龙教授拿出一本宋锦图册给小龙看,向小龙介绍宋锦的品种和纹样。

龙教授:小龙,你看看这本图册。

小　龙:哇,这些宋锦真精美呀,上面还标着织造时间和品名呢。

龙教授:你看这幅《彩织极乐世界图轴》,这是清代匠人们的杰作,是宋锦成熟期的作品。

小　龙:像画一样,真是令人难以置信!

龙教授:这幅宋锦作品其实就是描摹的一幅画作。据说乾隆皇帝曾请人为他母亲画了一幅《极乐世界图》,可是宫女不小心把画弄破了,他就让苏州织造署用宋锦复制了原画。

小　龙:原来是这样,现在这幅织锦应该比那幅画还有

《彩织极乐世界图轴》　*Pure Land Brocade Scroll*

名吧？

龙教授：的确是。这幅宋锦整个构图对称严谨，画面壮观，有278个神态各异的佛像，还有宫殿、祥云和奇花异草。整幅作品长约4.5米，宽约2米。

小　　龙：能织出来这么大的一幅作品，真了不起！

龙教授：确实是少见的大幅织锦画，国宝级文物。著名的宋锦传承人钱小萍老师用了整整6年的时间，才成功仿织出这幅作品。

小　　龙：看来织这样的画很不容易。爸爸，我看这图册里有很多式样，宋锦应该也分好多品种吧？

龙教授：是的，现在宋锦分四大类：重（zhòng）锦、细锦、匣锦[①]、小锦[②]。我们刚才看的《彩织极乐世界图轴》属于重锦，是宋锦中最名贵的品种。

小　　龙：那它的用料肯定特别高级、华贵。

龙教授：那当然了，使用金线是重锦的特色之一。你看，这幅锦画在人物头部和房屋的重点部位都用了金线。

小　　龙：还真是的，金色非常醒目。这么贵重，看来只有皇室和贵族才用得起。

龙教授：对。重锦主要用于宫廷装饰，锦面通常使用云龙纹样，象征吉祥和权力。

重锦　Exquisite Brocade　017

重锦　Exquisite Brocade

小　　龙：爸爸，我发现这幅锦画的色彩也很丰富。

龙教授：色彩丰富是重锦的另一大特色。就像这幅图轴，单单一个红色就分出大红、木红、粉红、水红等不同色阶。钱老师复制时用了26种不同颜色的纬线。

小　　龙：这么多种？

龙教授：对，重锦色彩丰富，织的时候需要多组纬线先后排列和上下叠加。

小　　龙：那经线应该没那么复杂吧？

龙教授：也比一般的织锦复杂，宋锦有两种经线。用来织底纹的经线叫作地经或底经；另外一种用来呈现锦面表层色彩和图案的叫面经，数量是底经的三分之一。面经采用比底经更细的白色生丝，用来压住表面的纬线浮长，使它们更牢固。

小　　龙：这里面两种经线分工明确，对吧？

龙教授：是的，更专业的叫法是双重（chóng）经线，使用双重经线的优点是宋锦正反面都很平滑。你再仔细看看、摸摸那条围巾。

小　　龙：嗯，正反面都很好看。既然织造重锦用到了多重纬线和双重经线，为什么不叫重（chóng）锦，而叫重（zhòng）锦呢？

龙教授：很多人还真这么叫呢，不过我认为，之所以叫重锦，主要是和细锦相对，也表示用了金银等贵重材料。

注释：
① 匣锦：用真丝与少量纱线混合织成的宋锦，图案连续对称，多用于画的立轴、屏风的装裱和礼品盒。
② 小锦：一种花纹细碎的宋锦，一般用于小件工艺品的包装盒。

Exquisite Brocade

> Prof. Long took out a picture book of Song brocade and continued to talk about the categories and patterns of Song brocade with Xiaolong.

Prof. Long: Look at this, Xiaolong.

Xiaolong: Wow! These brocades are truly works of art. And their names and completion time are listed here.

Prof. Long: Yes. Look at this one. It's a brocade scroll named *Pure Land Brocade Scroll*, a masterpiece created by weavers in the Qing Dynasty. This piece of work was produced in the period when the craftsmanship of Song brocade had matured considerably.

Xiaolong: A brocade scroll? Incredible! It just looks like a painting!

Prof. Long: Yes, it does. In fact, this piece of work was modelled on a painting. It's said that Emperor Qianlong once commissioned a painting titled *Pure Land* for his mother. Unfortunately, it was accidentally damaged by a palace maid. To remedy the situation, the emperor asked the Suzhou Weaving and Dyeing Bureau to reproduce this work with the weaving craft of Song brocade. That's how we got this brocade scroll.

Xiaolong: That makes sense. I guess the brocade scroll now is much more famous than the painting.

Prof. Long: Indeed. The brocade scroll is symmetrically designed with a magnificent spectacle of 278 varied Buddha statues as well as palaces, clouds of different colours, exotic flowers and herbs. It's about 4.5 metres long and 2 metres wide.

Xiaolong: What a huge piece!

Prof. Long: A brocade of this size is truly rare and definitely a national treasure. It took Ms. Qian Xiaoping, a well-known Song brocade inheritor, six full years to make a replica.

Xiaolong: It must have been very challenging! Dad, look at these Song brocades in the picture book. There are a lot of varieties. Does Song brocade have many categories?

Prof. Long: Yes. Song brocade actually falls into four major categories: exquisite brocade, fine brocade, box brocade[1] and small brocade[2]. The brocade scroll we have just seen is an exemplar of exquisite brocade, or *zhongjin* (重锦) in Chinese, the most precious kind of Song brocade.

Xiaolong: Well, the material of exquisite brocade must be of premium quality and costly then.

Prof. Long: Absolutely. A main feature of exquisite brocade is the use of gold threads. Look, in this brocade painting, the figures' heads and some parts of the buildings are all interwoven with gold threads.

Xiaolong: It really is. The gold colour is eye-catching. Only the royal and noble families can afford to use the valuable exquisite brocade, I guess?

Prof. Long: Of course. Exquisite brocade is mainly used for imperial decorations in ancient times. The cloud-and-dragon design is a typical example. It symbolises blessings and power.

Xiaolong: Dad, look at the brocade scroll. It also has many colours.

Prof. Long: That's another important feature of exquisite brocade, namely, the rich colours. As you can see in the scroll, a single red colour can be differentiated into various shades: bright red, mahogany red, pink, and bright pink. It is said that to make a replica of this scroll, Ms. Qian Xiaoping used 26 different colours of weft threads.

Xiaolong: That's quite a few.

Prof. Long: Yes. In order to create such a rich variety of colours for exquisite brocade, it requires the arrangement and overlapping of multiple weft threads.

Xiaolong: What about the warp threads then?

Prof. Long: The use of warp threads in exquisite brocade is more sophisticated than that in common brocades.

When weaving Song brocade, two types of warp threads are needed. They are respectively the base warp and the surface warp. The former is used to weave the ground pattern, and the latter, made of finer white raw silk, is to present the surface colours and patterns. The number of the surface warp is about one third of the base warp. It's also used to secure the weft threads, so that the texture would be tighter.

Xiaolong: Dad, you mean the two types of warp threads are used for different purposes.

Prof. Long: Right. Technically, the term is known as double warp threads. They combine to make both sides of the Song brocade smooth and lustrous. Xiaolong, you may take a closer look and feel the texture of that scarf.

Xiaolong: Well, both sides feel smooth and soft. I can't even tell which is the right side. That's amazing. Now I'd prefer to pronounce the Chinese name of exquisite brocade as *chongjin*, since the character *chong* (重)

could mean being double and multiple, which speaks of the important techniques used here.

Prof. Long: Actually, many people do pronounce the term in this way. But when pronounced as *zhongjin*, it would be more easily identified in a counterpart relation with fine brocade. The character *zhong* (重) is also suggestive of the use of precious materials like gold and silver threads.

Notes:

1. **Box brocade:** A variety of Song brocade made by mixing silk and a small quantity of yarn. It features the symmetrical design with repeating and continuous patterns. It's widely used for the purpose of decoration in hanging scrolls, mounted screens and gift packaging.

2. **Small brocade:** A variety of Song brocade with tiny patterns for the decoration of small craftworks.

细锦

> 小龙又仔细看看、摸了摸爸爸送给他的围巾,感觉双面平滑,而且轻薄柔软。

小　　龙：爸爸,这条围巾应该不是重锦,是细锦吧?

龙教授：没错。细锦是宋锦中最基础、最常见的一种。你应该也感受出来了,细锦质地柔软轻薄。

小　　龙：这和用线有什么关系呢?是因为用的线更细吗?

龙教授：对。织细锦用的丝线更细,通常也不用金线。还有一个关键点,就是细锦用的纬线重数没有重锦那么多,自然也不会那么厚。

小　　龙：您是说纬线重数越多,织锦就会越厚?

龙教授：通常是这样。不同颜色的经纬线层层交叠,才会将丰富的色彩呈现在锦面上。色彩的呈现主要是纬线的应用。如果要色彩丰富,往往得在某个纹样区域

细锦　Fine Brocade

里叠加不同颜色的纬线。这样一来，织出的锦自然也会厚一些。

小　龙：嗯，明白了。可是这条薄围巾颜色并不单调，还富有变化呢。这是怎么做到的呢？

龙教授：这用的是宋锦特有的一种技术。简单地讲，织完一组纹样后，直接换成其他颜色的纬线，就可以产生变换色彩的效果。

小　龙：噢，这样就不用为了增加颜色再额外叠加一组纬线了。

龙教授：没错。这叫分段换色，织一些重复出现的主题花纹时，会经常用到这个技术。

小　龙：我还真没认真想过这个问题，只是觉得颜色挺和谐的。现在才发现这条围巾的特点是花纹形状相同，颜色却不一样。

龙教授：用分段换色技术织出来的细锦既轻薄又好看，用作衣料再合适不过了。

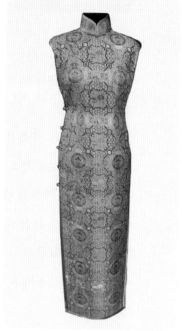

细锦服饰　Costumes of Fine Brocade

小　　龙：嗯，细锦做成的衣服穿着一定很舒适。

龙教授：实际上，宋代的时候，细锦就已经是官府制衣的主要面料。

小　　龙：细锦还有什么别的用途呢？

龙教授：除了衣料，细锦还常常用来装帧书籍、装裱高档书画等。过去，用细锦作装裱材料在士大夫阶层很流行，这也是刺激宋锦生产的一个原因。现在，宋锦的需求没有以前那么大，但常作为高级服饰和装裱材料使用。

Fine Brocade

> Xiaolong felt the texture of the scarf. It was smooth on both sides, light, thin and soft.

Xiaolong: Dad, this scarf is probably not made of exquisite brocade but rather fine brocade, right?

Prof. Long: Yes. Fine brocade is the most basic and common type in Song brocade. You see, it's quite soft and light.

Xiaolong: Is it because thinner threads are used in weaving fine brocade?

Prof. Long: You're quite right. Fine brocade is usually woven with thinner silk threads, not with gold threads. Also, the weft density of fine brocade is not as high as that of exquisite brocade.

Xiaolong: You mean that the higher the weft density, the thicker the brocade would be?

Prof. Long: That's true. The warp and weft of different colours are interwoven to form varied patterns and designs. The colour scheme mainly relies on the use of weft threads. If a vibrant colour effect is desired, it often needs multiple coloured weft threads overlaid within a specific pattern. In this case, the brocade gains thickness as well.

Xiaolong: Ah, I see. But this thin scarf has rich colours. How is that possible?

Prof. Long: This is achieved through a distinctive weaving technique specific to Song brocade. Simply put, to meet the need for colour variation, the weft threads in use can be replaced with another set of threads of different colours.

Xiaolong: Oh, there's no need to add up a new set of weft threads then.

Prof. Long: Exactly. It's called segmented colour change. This technique is often used in weaving recurrent motif patterns when there's a need for colour variation.

Xiaolong: That's cool. Now I see the ingenuity behind the harmonious combination of colours.

Prof. Long: Fine brocade made in this way is light and beautiful, which is perfect for making clothes.

Xiaolong: Oh, the clothes made of brocade must be very comfortable.

Prof. Long: As a matter of fact, during the Song Dynasty, fine brocade was already one of the designated clothing fabrics for officials.

Xiaolong: Are there any other uses for it?

Prof. Long: On top of that, fine brocade was often used for binding books and framing artworks of calligraphy and paintings. This was especially popular and in great demand among scholar-officials in the Song Dynasty, which promoted the production and sales of Song brocade. Nowadays, people's demand for Song brocade may not be as great as it used to be, but it has always been a popular choice for high-grade clothing and artwork decorations.

几何纹

龙教授：小龙，你看这条围巾上面是什么纹样？

小　龙：有好多菱形花纹。

龙教授：这是一种比较简单的几何纹样。几何纹是宋锦中常见的纹样，由一些基础的几何图形变化而来。

小　龙：这些菱形花纹对称规整，而且是连续排列的。

龙教授：对，正是因为这种连续排列，用分段换色正好可以突出色彩的变换。这种排列其实也有源远流长的寓意。宋朝文化讲究严谨雅致，从这些纹样上就能看出来。

小　龙：宋锦有哪些典型的几何纹样呢？

龙教授：像球路纹、龟背纹、八达晕纹，这些都是宋锦中比较常见的几何纹。

小　龙：听上去球路纹应该是一些圆形纹样吧？

龙教授：对，球路纹是一种大圆和小圆相切的图案布局，取

几何纹　Geometric Design

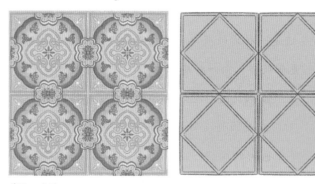

菱格四合纹　Diamond-Shaped Pattern

球路纹　Sphere-Shaped Pattern

"珠联璧合"的美好寓意。宋锦非常特别的一点，就是把中国人追求美好的心理和意识表现得既含蓄又文雅，人们一看纹样就能心领神会。

小　　龙：这些几何纹样不仅好看，而且还深藏寓意呢。

龙教授：没错，你来猜猜龟背纹的寓意。

小　　龙：龟背纹应该就像龟壳上的图案那样，是连续的六边形吧？乌龟长寿，龟背纹有健康长寿的意思吧？

龙教授：非常正确。再猜猜八达晕纹？

小　　龙：八达……是"四通八达"的意思吗？

龟背纹　Turtle Shell Pattern

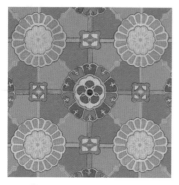 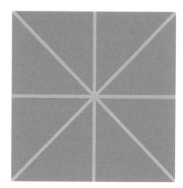

八达晕纹　Eight-Direction Radial Pattern

鸳鸯瑞花纹宋锦　Song Brocade with Mandarin Duck Pattern

龙教授：没错。"八"表示八个方位。八达晕纹以"米"字格为基本骨架，线与线相交，朝八方辐射。这个纹样喻示着事事顺利、畅通无阻。

小　龙：这里的"晕"又是什么意思？

龙教授："晕"是指循序变化的色彩效果，体现的是织锦色彩的层次感，比如从深黄色过渡到浅黄色。实际上，除了八达晕纹，还有四达晕纹和六达晕纹。

小　龙：是不是说这一系列纹样都有类似的色彩效果，只是几何纹中的线条数不一样？

龙教授：对。在线条相交处多用方形、圆形和多边形，形成整个锦面图案的框架，然后在几何形内外空间中填入各种小的几何纹和折枝花纹等，都是各种寓意吉祥的元素。

Geometric Design

Prof. Long: Xiaolong, did you notice the patterns on this scarf?

Xiaolong: You mean the diamond-shaped pattern like this one? There're quite many actually.

Prof. Long: Yes. It is a simple form of geometric design. Geometric patterns are commonly seen in Song brocade and they're derived from basic geometric shapes.

Xiaolong: I see. I also noticed that these diamond-shaped patterns are symmetrical and aligned in succession.

Prof. Long: It's true. Against this geometric regularity, the segmented colour variation would easily stand out. Such continuous and seamless designs actually imply vitality and sustainability. The aesthetic principles of neatness and elegance highly valued in the Song Dynasty are graphically expressed here.

Xiaolong: So it's a combination of wit and style. What are the typical geometric patterns used in Song brocade?

Prof. Long: The most common forms include sphere-shaped pattern, turtle shell pattern, eight-direction radial pattern, to name but a few.

Xiaolong: There should be many round-shaped figures on the fabric with the sphere-shaped pattern.

Prof. Long: You're right. You'd find a structure consisting of circles of different sizes. The design embodies the auspicious concept of "harmonious union". You see, what is remarkable about Song brocade is its artistic expression of the Chinese perspective on beauty. You could tell that these patterns also display sentiments of subtlety.

Xiaolong: Indeed. These patterns are aesthetically pleasing and rich in cultural connotations.

Prof. Long: All right. Let's come to the turtle shell pattern. Can you guess what the implicit meaning is?

Xiaolong: Let me see. Look at these seamless hexagons. They just look like the turtle's shell. Turtles live a long life, so the pattern means health and longevity, right?

Prof. Long: Exactly! What about *badayun*（八达晕）, or eight-direction radial pattern then?

Xiaolong: Hmm, it reminds me of the Chinese idiom *sitong bada*（四通八达）meaning "easy access from all directions". Does it have something to do with it?

Prof. Long: Yes. The character *ba*（八）, or eight, here signifies the eight directions which in Chinese stand for the whole space. The pattern basically resembles the form of the Chinese character "米" with four intersecting lines splitting the space into eight directions. It implies everything goes smoothly without obstacles.

Xiaolong: I see. Then what about the Chinese character *yun*（晕）in the name of this design?

Prof. Long: It refers to the gradual change of colours, such as the transition from dark yellow to light yellow. The gradation of colours on the brocade always makes a visual impact. Besides eight-direction radial pattern, there are also four-direction radial pattern and six-direction radial pattern.

Xiaolong: As their names could tell, they should have the similar designing feature but the number of radial lines in the pattern is different.

Prof. Long: You're right. Usually at the intersection of these lines, we'd also see some geometric shapes, like squares, circles, polygons. They constitute the frame of the pattern. There, it'll be further filled with smaller geometric designs and some auspicious elements.

天华锦

> 小龙和爸爸继续讨论宋锦的图案纹样。

小　　龙：爸爸，这样看来，宋锦中几何纹不仅起到装饰作用，同时还是锦面纹样的骨架。

龙教授：是的，几何纹决定着整个图案布局，是宋锦锦面的基本纹样，之后还需要锦上添花。

小　　龙：就是再添上花纹的意思吗？

龙教授：字面上可以这么理解。这里的"花"并不单指花朵纹样，而是指各种图案，涉及花卉、动物、器物等许多题材。

小　　龙："锦上添花"就是在几何底纹的基础上，再添加精美的图案？

龙教授：是的。"锦上添花"其实是用到了一种创造性的纹样设计方法——叠加。

小　龙：叠加上去的图案是不是也很有讲究？

龙教授：对，这些图案都有美好寓意。比如，莲花纹指代出淤泥而不染的高尚品质，狮子纹寓意喜乐吉祥，五个蝙蝠暗合五福[①]圆满等等。还有一些图案与宗教有关。

小　龙：与宗教有关的图案有哪些呢？

龙教授：道教和佛教都对宋代的审美产生过深远的影响。在宋锦上会用到一些道教法器图案，比如宝剑和宝镜，还有佛教宝器图案，比如花伞、法轮和双鱼。

小　龙：看来宗教信仰也会影响当时的纹样设计。

龙教授：没错。最能体现锦上添花高超技艺的是天华锦。

小　龙：咦？听上去就像"添花"。

龙教授：嗯，其实天华锦也叫"添花锦"，通过谐音取"锦上添花"的意思。

小　龙：这样一下子就能记住"锦上添花"这个成语，太形

锦上添花
Adding Exquisite Patterns to Brocade

象了。

龙教授：是的。这种锦的纹样非常精美，蜀锦和云锦里也都有天华锦这个品种。天华锦的底纹比较特别，通常由圆、方、菱形、六角形、八角形等几何形状交错排列，组成富有变化规律的纹路骨架。由这些骨架分割出的区域有主次之分，在上面添的图案也不一样。

小　龙：是要突出主题纹样吗？

龙教授：是的。主体区域通常会用比较大的主题纹样，其他区域里的纹样起衬托作用。

小　龙：为什么要这样设计呢？

龙教授：这样能形成一种锦纹多样、变化丰富，但又和谐统一的风格。这也是宋锦的最大特色。就拿故宫收藏的"红色地方棋朵花四合如意纹天华锦"来说……

小　龙：这名字好长呀。

黄地龙凤如意八达晕天华锦
Yellow-Ground *Tianhua* Brocade of Eight-Direction Radial Patterns with Chinese Dragon and Phoenix Patterns

蓝地龙凤纹天华锦
Blue-Ground *Tianhua* Brocade with Chinese Dragon and Phoenix Patterns

龙教授：确实很长。这里的"红色地"说的是锦面的底色；"方棋"指的是由方棋格构成纵横交错的几何形骨架；"朵花"是中间菱形里的团形主题图案，辅助的是"四合如意"云纹。

小　龙：嗯，这么一解释，我就明白了。名字是长了点，但意思很清楚。

龙教授：对。这幅锦有团圆如意、天下四合的美好寓意。你再看看这段龙凤呈祥的织锦，里面的纹样同样也有主次之分，凤纹和宝相花属于辅纹。

小　龙：那上面的云龙纹应该就是主题图案吧。

龙教授：对。你有没有注意到主花色调与底色之间有反差？

小　龙：这也是凸显主题纹样的一种方式吗？

龙教授：没错。适当的反差使整个锦面层次分明，突出龙凤呈祥的寓意。

注释：

① 五福：指长寿、富贵、康宁、好德和善终。五福概念最初出现在儒家经典《尚书》里。

Tianhua Brocade

> Xiaolong and Prof. Long continued talking about the patterns of Song brocade.

Xiaolong: Dad, now I see that geometric patterns not only serve a decorative purpose. Their structural functions are prominent as well.

Prof. Long: Correct. Geometric patterns stand as the basic structure of the whole design and determine the overall layout. Then, we also need to add up additional patterns, or *tianhua* (添花) in Chinese.

Xiaolong: Does the term mean adding floral patterns to the brocade?

Prof. Long: Well, you can take that literally. But *hua* here is not confined to floral patterns. It could feature plants, animals, wares, among many others.

Xiaolong: Oh, I see. *Tianhua* means adding exquisite patterns to the basic geometric patterns. Am I right?

Prof. Long: Yes. This technique needs a creative method called superimposition.

Xiaolong: Okay then. Do the superimposed patterns mean anything special?

Prof. Long: Yes. These patterns usually have good implications. For example, the lotus pattern symbolises pure and untainted nature, the lion pattern signifies happiness and blessings, the five-*fu* (蝠 , bat) pattern represents the five blessings[1] to the full. What's more, some patterns are also related to religion.

Xiaolong: Could you tell me a bit more, Dad?

Prof. Long: Sure. Both Taoism and Buddhism had a profound influence on the aesthetic values in the Song Dynasty. It could be well reflected on Song brocade. For example, you can find some patterns of Taoist artefacts like swords and mirrors and Buddhist instruments like the umbrella, the wheel of the law, and twin fish.

Xiaolong: Well, the creation of patterns on Song brocade also bears some socio-cultural imprints.

Prof. Long: Sure. Back to the superimposition of patterns. There's actually a type of brocade that could fully demonstrate the skill of superimposition. It is known as *tianhua* (天华) brocade.

Xiaolong: It rings a bell. It sounds quite similar to the term *tianhua* (添花), meaning adding up exquisite patterns.

Prof. Long: Yes. Sometimes it is pronounced in this way, indicative of the Chinese idiom adding flowers on the brocade, or *jinshang tianhua* (锦上添花) in Chinese. It means making what is good even better.

Xiaolong: What a catchy phrase!

Prof. Long: *Tianhua* brocade features incredibly exquisite patterns. It is not exclusive to Song brocade though. Shu brocade and Nanjing Yunjin brocade also boast this type of design. One of the distinctive features of *tianhua* brocade lies in its ground pattern. It typically consists of a geometric arrangement of circles, squares, diamonds, hexagons, octagons, or other shapes, forming a structure with regular variations. Some parts on the fabric are sectioned off as the central area for superimposition while the rest the subsidiary.

Xiaolong: The central area is of greater importance in design, right?

Prof. Long: Absolutely! Usually, larger theme patterns would be superimposed in the central area. The remaining parts are filled with supplementary elements.

Xiaolong: What's that for?

Prof. Long: That will help to increase the richness of the design while maintaining a harmonious style. Diversity plus unity, this is actually the signature of Song brocade. Take this piece from the Imperial Palace as an example. It's called "red-ground chequered *tianhua* brocade with floral design and quadruple *ruyi*（如意）pattern".

Xiaolong: Such a long name.

Prof. Long: Indeed. As its name could tell, the background

colour of the brocade is red. It features a chequered design so you could see a criss-cross geometric frame composed of squares. You could also find some beautiful floral images in the central area surrounded by *ruyi* patterns.

Xiaolong: That explains a lot. Although it has a long name, its meaning is quite clear.

Prof. Long: You're right. This brocade expresses the good wishes of harmony, unity, and prosperity. Now, let's come to this Chinese dragon-and-phoenix brocade. There're also theme and subsidiary patterns when it comes to the design. Look, the image of phoenix and coiled flowers are the subsidiary patterns.

Xiaolong: Oh, I see! Then the cloud-and-dragon design must be the theme pattern.

Prof. Long: You're right. Have you noticed the contrast between the colours of the main patterns and the background?

Xiaolong: Yes. I assume this is to highlight the theme patterns, right?

Prof. Long: Exactly. In this way, the multiple layers of the brocade will become clearer. It'll also help to express the theme of prosperity as the patterns of Chinese dragon and phoenix could imply.

Note:

1. Five blessings: They refer to longevity, wealth, health, virtue and a peaceful death. The concept of five blessings first appeared in the Confucian classic *Book of Documents*.

灯笼锦

小　　龙：爸爸，宋锦的命名很能体现古人的审美趣味，文化内涵真是丰富。

龙教授：确实，古人给织锦命名常会蕴含美好的寓意。比如，"灯笼锦"就是很典型的例子。

小　　龙："灯笼锦"？这个名字很有趣。上面肯定是织了灯笼的图案！

龙教授：对，灯笼是主要图案。你想想看，锦面上的灯笼图案并行排列，是不是就像元宵节华灯齐放的景象？

小　　龙：嗯，很容易让人联想到满城灯火、欢喜过节的场景，很有生活气息。

龙教授：唐宋时期，正月十五是个很重要的节日，灯节的晚上，花灯千姿百态，熠熠生辉。灯笼锦取的正是这个意象。人们也把灯笼锦叫"天下乐锦"，就是取"元宵灯节、君民同乐"的意思。

灯笼锦　Lantern Brocade

小　　龙：看来灯笼锦同样也是借助纹样传达美好的寓意。

龙教授：对。灯笼锦一直流行到明清时期。这种宋锦上面的灯笼纹，其实是取自臣僚袄子锦中的一种纹样。

小　　龙：臣僚袄子锦是什么？

龙教授：宋朝皇帝每年端午节和十月初一会赏给百官锦缎。按照官品不同，所赐锦缎上的纹饰有所不同。官员们会把皇帝赏赐的锦缎缝在官袍的前胸和后背上，这就是臣僚袄子锦的来历。

五谷丰登纹灯笼锦　Good-Harvest Lantern Brocade

小　　龙：那不就是清朝官服上的补子吗？

龙教授：是的，后来慢慢就演变成明清官服上的补子了。

小　　龙：没想到补子还有这么个来历。

龙教授：时间久了，灯笼纹也演变出不同的样式，锦面上用来装饰灯笼图案的元素也丰富起来。

小　　龙：那寓意就更丰富了。

龙教授：是的。比较经典的纹样是在灯笼四角配上流苏。流

吉庆有余纹灯笼锦　Full-Festivity Lantern Brocade

苏一般用的是谷穗图案。灯的周围还会织上飞舞的蜜蜂。

小　　龙：该不会是五谷丰登的意思吧？

龙教授：哈哈，我儿子很聪明啊，一下就猜到了。

小　　龙：还有没有其他的样式？

龙教授：有啊。有的灯笼锦的灯壁上垂挂着吊珠……

小　　龙：意思是珠联璧合吗？

龙教授：是的。还有一种是在灯下悬坠石磬[①]和玉鱼。

小　龙：那一定是吉庆有余了。看来灯笼锦的纹样设计里面藏着不少文化意象。

龙教授：是啊，要明白纹样的意思，得要联想到这些图案元素的谐音。

注释：
① 石磬：简称"磬"，一种板形击奏乐器，是中国古代礼乐中重要的乐器。

Lantern Brocade

Xiaolong: Dad, I found that the patterns on Song brocade not only reveal the aesthetic values of ancient Chinese people but also embody rich cultural connotations.

Prof. Long: That's true. It's also quite interesting to see that in ancient China, a brocade was often named after rich blessings. Lantern brocade is a very typical example.

Xiaolong: There must be lantern designs on the brocade.

Prof. Long: You're right. Lanterns make up the primary pattern of this brocade. Just imagine two parallel lines of lantern designs. Don't they look like the beautiful scene of the Lantern Festival with all the lights on?

Xiaolong: Yes. They remind me of the festive vibe.

Prof. Long: Quite true. During the Tang and Song dynasties, the Lantern Festival on the fifteenth day of the first lunar month was a very important occasion. On that day, lamps and lanterns of different shapes shone brightly at night. That's where the name "lantern brocade" comes from. At that time, people also called it "brocade of happiness under heaven", meaning that the ruler and the people shared the joy on the Lantern Festival.

Xiaolong: I see. Like other brocades, the pattern of lantern brocade also expresses good wishes.

Prof. Long: Yes. The popularity of lantern brocade continued unabated until the Ming and Qing dynasties. And the design of lantern patterns actually has its origin in a particular brocade found on officials' robes in ancient China.

Xiaolong: Are there anything special about this particular type of brocade?

Prof. Long: Of course. It's a kind of brocade bestowed upon officials by the emperor in the Song Dynasty. It usually happened on Duanwu, namely, the fifth day of the fifth month, and the first day of the tenth month according to the Chinese lunar calendar every year. The patterns on the brocade would vary with the official rank. The brocade would be sewed at the front and back of officials' robes.

Xiaolong: Like the badges on officials' robes in the Qing Dynasty?

Prof. Long: Exactly. Over time, they gradually evolved into the decorative badges seen on Ming and Qing officials' robes.

Xiaolong: That's quite interesting.

Prof. Long: Well, as time went by, varied lantern patterns had been developed and the elements on lantern brocade were also enriched.

Xiaolong: That means more cultural meanings.

Prof. Long: That's true. On the classic design, you can find the element of tassels on the four corners of the lantern, resembling the ears of grain. Around the lantern there're two dancing bees.

Xiaolong: Does this mean a good harvest?

Prof. Long: Haha, that's my boy. Correct!

Xiaolong: So interesting! Are there any other styles?

Prof. Long: Of course. There are different varieties. For example, some designs have strings of pearls on the lantern tassels.

Xiaolong: Oh, that reminds me of the Chinese idiom "strings of pearls and fine jades in one piece", which usually symbolises a perfect match.

Prof. Long: That's exactly the message the lantern design intends to convey. There's another design worth mentioning. At the lower part of the lantern, you can see the image of the stone percussion instrument[1], or *qing* in Chinese, and the jade in the shape of fish.

Xiaolong: Wait. Fish in Chinese is pronounced as *yu*. *Qing* and *yu*, these two characters allude to the Chinese idiom "*jiqing youyu*"（吉庆有余）, meaning overabundance of auspiciousness and happiness. What an ingenious design to perfectly blend all these cultural images!

Prof. Long: I couldn't agree more. And one of the effective ways to figure out the underlying meanings is to refer to their pronunciations and homophones in Chinese.

Note:

1. **Stone percussion instrument:** *Shiqing*（石磬）in Chinese or *qing*（磬）for short. It played a prominent role in the ancient ritual and music culture in China.

福娃纹

> 龙教授翻到2008年北京奥运会的福娃邮票纹样,指给小龙看。

龙教授:小龙,你看,这是什么?

小　龙:这不是邮票嘛。

福娃纹
Fuwa Design

龙教授：再仔细瞧瞧。

小　龙：是宋锦做的邮票吧？

龙教授：猜对了。这是为2008年北京奥运会特别制作的福娃①纹宋锦邮票。这可不只是纪念品，也是真正可以使用的邮票，是当时有个国家专门定制的，还在

福娃纹宋锦邮票　Song Brocade Stamp with Fuwa Design

全世界发行了呢。

小　　龙：邮票那么薄，那么小，上面还得有图案、文字，能织出来可真了不起。

龙教授：想不到吧。你看，宋锦质地轻薄平整，而且注重各种颜色、纯度和明度的协调统一，做出的邮票也很漂亮。

小　　龙：这在全世界应该是独一无二的吧。

龙教授：当然了。把宋锦织造技艺和邮票设计结合起来，无论是在世界邮票史上，还是宋锦织造史上，都是前所未有的。当时钱小平老师花费半年多的时间不断研究和实验，才设计出这款织锦邮票的结构组织，成功地把奥运五环和福娃纹都织在了这方寸邮票上。

小　　龙：这织出来的福娃好可爱呀。

龙教授：你看，奥运五环和福娃的颜色相呼应，五环代表奥运精神，五个福娃表达了"北京欢迎你"的热情，把它们织在一起，意义非凡。

小　　龙：这里面是不是还有别的寓意呀？

龙教授：是啊，这套福娃纹宋锦邮票其实还有"锦绣奥运"、"锦绣中国"和"锦绣世界"的含义，借此向世界表达美好愿景。

小　龙：明白了，这是借用了汉语里"锦绣"二字，因为它们除了指精美的丝织品和绣品，还用来比喻美好的事物。真是与时俱进！

龙教授：理解得很正确。这可以说是宋锦织造传承和创新的杰作。新中国成立前，宋锦已濒临绝迹。新中国成立后，宋锦织造业才得以复苏。如今，我们不仅复原了一些传统纹样，还不断推出富有时代特色的新纹样。能有这样的成绩，我们可不能忘了非遗传承人的工匠精神啊。

注释：

① 福娃：2008年北京奥运会吉祥物，共有五个，分别叫贝贝、晶晶、欢欢、迎迎和妮妮。它们的名字单字连起来，同"北京欢迎你"谐音。

Fuwa Design

> Prof. Long showed Xiaolong a set of stamps made of Song brocade with the image of Fuwa mascots on the fabric.

Prof. Long: Hey, Xiaolong, look at this.

Xiaolong: Oh, it's a sheet of stamps.

Prof. Long: Are you sure? You'd better take a closer look.

Xiaolong: What? Don't tell me it's made of Song brocade!

Prof. Long: It really is! It's Song Brocade Stamp with Fuwa[1] design tailor-made for the 2008 Summer Olympics in Beijing. Don't simply take it as a souvenir. It's a genuinely usable postage stamp officially commissioned at that time in 2008. Later, it was issued all over the world.

Xiaolong: Incredible! It must be very difficult to weave these patterns and characters on Song brocade.

Prof. Long: Surprising, isn't it? As you can see, Song brocade has a light, smooth texture and its weaving techniques could guarantee a harmonious coordination of different colours and shades. It's a perfect material for making exquisite stamps.

Xiaolong: And there's no precedent for this.

Prof. Long: No. Back then, Ms. Qian Xiaoping, the renowned Song brocade master, took over half a year to research and experiment on the weaving of the brocade stamps. She finally managed to weave the patterns of Olympic rings and Fuwa mascots on this small-sized fabric.

Xiaolong: Really adorable!

Prof. Long: The colour of the Olympic rings and that of the Fuwa are an exact match. The five rings symbolise the spirit of the Olympics, while the five Fuwa mascots convey the message of "Beijing welcomes you".

Xiaolong: The design is of great significance.

Prof. Long: Actually, these Song brocade stamps also convey the meanings of "wonderful Olympic Games", "fantastic China" and "splendid world", expressing good wishes to everyone.

Xiaolong: I see. It draws inspiration from the Chinese term *jinxiu* (锦绣) which not only literally refers to exquisite silk fabrics and embroideries but also symbolises all the wonderful things. It truly keeps up with the times!

Prof. Long: You're quite right. This product can be regarded as a representative work in the drive to inherit and innovate the skills of Song brocade weaving at contemporary times. Song brocade industry was once on the verge of extinction, and after the People's Republic of China was founded in 1949, this industry revived. Today, we not only successfully bring some traditional patterns back to life, but also introduce some new designs that could reflect the spirit of the times. And these achievements are credited to the dedicated practitioners in the fields of Intangible Cultural Heritage.

Note:

1.Fuwa: The mascots of the 2008 Summer Olympics in Beijing. There are five figures: Beibei, Jingjing, Huanhuan, Yingying, Nini. The monosyllabic character of their Chinese names combine to form the sentence "Beijing huanying ni", which means "Beijing welcomes you".

海水江崖纹

> 龙教授把图册翻到最后一页,是一张2014年亚太经济合作组织领导人非正式会议集体照。

龙教授:小龙,说起宋锦的传承和创新,必须提一下2014年的亚太经济合作组织领导人非正式会议。

小　龙:就是咱们作为东道主的那次?

龙教授:对。这个会议一直有个不成文的规定,就是由东道主提供样式统一的休闲服装,而且要体现出举办国的文化传统。

小　龙:是不是那次会议用宋锦做了礼服?

龙教授:是的。那次会议最后选用了宋锦做衣料,制成了新中式礼服。宋锦具有亚光特性,华而不炫、贵而不显,与我们含蓄内敛的文化气韵非常吻合。

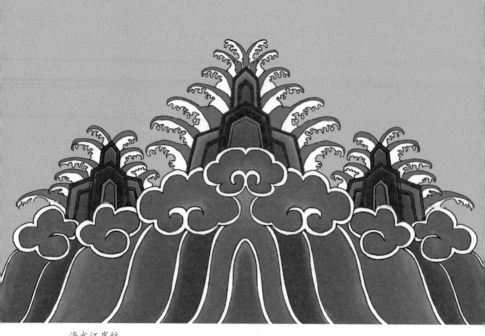

海水江崖纹
Sea-and-Mountain Pattern

小　　龙：宋锦制成的礼服，上身效果一定非常棒吧？

龙教授：设计师在面料上作了大胆改良。双经线还是使用蚕丝线，但把纬线改用羊毛纤维，这样做出来的礼服挺括有型。

小　　龙：那应该算是宋锦在传承中的又一次创新尝试吧？

龙教授：对。礼服上织有经典的海水江崖纹，精美的线条和色彩，展现出内敛而庄重的整体效果。

小　　龙：海水江崖纹？这好像是个传统纹样，我记得在京剧戏服上见过。

龙教授：这是个非常古老的纹样，常在瓷器、漆器、石雕、砚台上出现，但用在服装上只有六七百年的历史，是龙袍、官服下摆和袖口上常用的吉祥纹样。

小　龙：这套衣服上选用这种纹样有什么寓意呢？

龙教授：我们先看看这个纹样本身的含义。这个纹样里有重叠的山头，也就是江崖纹。"江崖"二字与"姜芽"谐音，姜芽生长茂盛，生命力强，因而江崖纹象征着万世升平、江山永固。

小　龙：就是希望国家永远和平、人民生活永远幸福的意思，对吧？

龙教授：对。海水江崖纹下端是斜向排列的曲线，称为"水脚"。水脚上除了有挺立的山石，还有翻卷的海浪。这些水纹代指大江大河，象征着国家和民族。

小　龙：明白了，这一套服装太有创意了。

龙教授：更深层的意思是希望参加会议的国家山水相依、守望相助。当时选用了五六种色调供与会领导人选择。服装的式样一致，但色调不一，传递的正是中国人"和而不同"的理念，同时寄托了中国人民与世界各国建立和谐友好关系的美好愿望。

小　龙：这里面学问还真不少。

龙教授：是的，就说色彩搭配吧，海水江崖纹的色彩主要分为单色和复色两大类。过去单色的海水江崖纹较少，多见于红色和蓝色龙袍。

小　龙：这次的服装好像是单色的海水江崖纹。

龙教授：还不完全是，严格地讲，男装是低调的黑紫两色。女装是蓝色同色系复色。

小　龙：怪不得不像戏曲服装那么张扬，原来是色彩选择的原因。

龙教授：是的，常规复色海水江崖纹颜色夸张，就是为了张扬权威，凸显地位。

小　龙：那复色有什么搭配规律呢？

龙教授：是根据水平纹样分层进行配色，色彩之间会有一定的渐变规律。复色海水江崖纹的主题色调，会和面料的整体颜色保持统一。比如，石青色衣服上的海水江崖纹多以蓝色为主，配上醒目的红色和金色；明黄色龙袍多以金色、红色为主，搭配蓝色或绿色。

小　龙：爸爸，光听宋锦的种类和纹样，就觉得技艺高超，又特别有文化含义。我想邀请大卫一起去苏州丝绸博物馆看看，长长见识。他可是个十足的中国文化迷呢。

龙教授：哦，你那个英国朋友大卫，应该带他去看看。苏州丝绸博物馆里面的宝贝可不少。你们到了博物馆之后，可以去请教我的好朋友王老师，让他给你们讲讲宋锦的织造技艺和历史。

小　龙：好！

Sea-and-Mountain Pattern

> Prof. Long turned to the last page of the Song brocade picture book. It was a photo of the 22nd APEC Informal Leaders' Meeting.

Prof. Long: Xiaolong, speaking of the inheritance and innovation of Song brocade at contemporary times, there's an occasion that is worth mentioning, namely, the 22nd APEC Informal Leaders' Meeting in 2014.

Xiaolong: In my memory, it was held in Beijing, right?

Prof. Long: Yes. Usually, the host country would design outfits of a distinctive local style for all the guests.

Xiaolong: I see. So did the guests all wear Song brocade-made outfits at that meeting?

Prof. Long: Exactly. Song brocade was chosen as the clothing material. The outfit adopted the neo-Chinese style. The material of Song brocade is smooth but not over-shiny, precious but not flashy. It well presents the quality much cherished in traditional Chinese culture.

Xiaolong: Exactly. I guess the Song brocade clothing must fit well.

Prof. Long: Indeed. It's worth mentioning that the designers of the Song brocade outfit have made some bold efforts in innovating the fabric. Wool was used for weft and silk for double warp. The costumes made of this material do look elegant and well-shaped.

Xiaolong: Oh, I believe this marks another attempt to make innovative use of Song brocade in its inheritance.

Prof. Long: Points taken. The outfit also featured the classic sea-and-mountain pattern with exquisite designs and colours exhibiting a restrained sense of dignity.

Xiaolong: The sea-and-mountain pattern? Is it also a traditional pattern? I think I've seen some similar pattern on Jingju costumes.

Prof. Long: Yes, it is an age-old pattern. You can find it on ancient porcelain ware, lacquerware, stone carvings, and inkstones. Its application in costumes has a history of 600 to 700 years. As an auspicious pattern, it was commonly seen on the robes of the emperors as well as on the hem and the sleeves of officials' robes in ancient China.

Xiaolong: I see. What's the implied meaning of this design?

Prof. Long: Well, let's take a look at its name first. The Chinese term *jiangya* (江崖) literally refers to the mountains in endless succession, representing the country's landscape and territory. It has the same pronunciation with *jiangya* (姜牙), or ginger buds that can grow luxuriantly with exuberant vitality. Hence, this pattern gives the blessings for national prosperity and stability.

Xiaolong: So, it conveys the wish for the country to be peaceful and the people to live happily, right?

Prof. Long: Yes. At the bottom of this pattern, you can find some curved lines inclining to the ground. They are figuratively called "water feet". On these water feet stand firm rocks amid rolling waves. These elements symbolise the life and vitality of the whole nation.

Xiaolong: The design is really special.

Prof. Long: The deeper meaning here is the hope for the participating countries to be interconnected and mutually supportive. There were also some different clothing colour choices available. While the style of these costumes was consistent, the colours varied. This conveys the Chinese concept of "harmony in diversity" and expresses the Chinese people's aspiration to establish harmonious and friendly relations with other countries around the world.

Xiaolong: There's so much wisdom behind it.

Prof. Long: Indeed. Usually the colour scheme in the sea-and-mountain pattern falls into two types, namely, monocolour and multicolour. In the past, the monocolour design was less common and mostly seen on emperors' robes with the base colour of red or blue.

Xiaolong: I remember the APEC costumes adopted the single-colour scheme.

Prof. Long: But not entirely in the strict sense. Actually, the men's costume featured a subtle mixture of black and purple while its female counterpart combines different shades of blue.

Xiaolong: No wonder the outfits don't look as showy as traditional opera costumes. It's all thanks to the wise choice of colours.

Prof. Long: Exactly. Traditional sea-and-mountain patterns like those you noticed on Jingju costumes are often bold in colour scheme, with the purpose to convey a magisterial sense.

Xiaolong: I see. Are there any specific rules behind the colour scheme especially when it comes to multicolour design?

Prof. Long: Sure. For instance, the pattern could be theoretically divided into different sections on the horizontal plane, which make the gradation of colour possible.

Also, the thematic colour of the design should be harmonised with the overall tone of the fabric. Like on a garment with a midnight blue colour, the pattern is mainly in blue, accompanied by vibrant hues of red and gold, while on a bright yellow Chinese dragon robe, the dominant colours are gold and red, complemented by shades of blue or green with cold undertones.

Xiaolong: That's a bit complicated. The craftsmanship of Song brocade really needs further study. Now I've learnt some of its varieties and patterns as well as the rich cultural connotations behind. I bet there're more to come. Dad, I want to invite David to visit Suzhou Silk Museum. He is a big fan of Chinese culture.

Prof. Long: David? Oh, your friend from UK, right? That's a good idea. There're many treasures in Suzhou Silk Museum. When you are there, you could visit my friend Mr. Wang and learn about the skills and history of Song brocade weaving.

Xiaolong: That's great!

脚踏缫丝车

> 小龙和大卫到苏州丝绸博物馆参观,他们在前台见到了龙教授的朋友王老师。

小龙、大卫:王老师好!

王老师:小龙,大卫,欢迎,欢迎!龙教授让我给你们科普一下宋锦知识,你们都想知道些什么呢?

小　龙:您先给我们讲讲宋锦织造的主要工序,好吗?

王老师:好的。织造宋锦的工序非常复杂,从缫丝到成品前后有20多道工序。我们边看边说吧。

大　卫:缫丝?我记得我们在蚕桑实习基地参观时,那里的技术员说过,缫丝就是把蚕丝从蚕茧上抽出来。

王老师:没错,把蚕丝抽出来之后,再合并成丝线,这也是织锦的第一步。你们知道一只蚕茧可以抽出多长的丝吗?

脚踏缫丝车
Treadle Reeling Machine

大　卫：那么小一个蚕茧，应该不会很长吧？

王老师：恰恰相反，可以抽出来很长的丝线。春茧茧丝一般长900～1500米，夏秋茧长700～1200米。

大　卫：没想到这么长！还真不能小看这小小的蚕茧呀。

小　　龙：我记得缫丝这道工序挺复杂，需要煮茧、索绪等等，还要用到专门的工具。

王老师：看来你们都已经很内行了。缫丝得用到一件重要的工具，叫缫丝车。

小　　龙：这个之前没听说过。

王老师：你们看，这就是我们按照历史文献仿造的脚踏缫丝车。

大　　卫：需要用单脚操作？

王老师：对。其实最原始的缫丝工具就是盆和筐，人们将蚕茧浸泡在热水盆中，用手抽丝，把丝线绕在丝筐上。古时候因为生产力低下，缫丝技术的发展一直非常缓慢。经过千百年的不断改进，到唐朝人们才开始广泛使用手摇缫丝车。操作时，人们一边煮茧，挑出每个茧上的丝头；一边操纵摇杆，带动一个装置，把引出来的蚕丝缠绕在上面。

小　　龙：就是说要一边煮茧一边缫丝，感觉两只手一直在忙活呢。

王老师：对。不光自己忙，还需要其他人在旁边添茧、烧火。

小　龙：那可得控制好速度，万一放的蚕茧多了，来不及操作，就会煮过头了吧？

王老师：一点没错。你们肯定知道，煮茧是为了借助热水软化蚕丝外层的丝胶，使牢牢粘在一起的蚕丝散解开，所以缫丝速度必须得快，以防丝胶过度溶解。

大　卫：是不是应该在煮茧环节上优化一下？

王老师：你说到点子上了。宋代后期，南方地区出现了冷盆缫丝法，就是先将茧煮好，放入温水中冷却，然后再慢慢进行缫丝。

大　卫：这是个好办法。

王老师：是的，通过这种办法可以更好地控制蚕丝的粗细，抽出来的丝也更有光泽、更有韧性。

小　龙：什么时候开始用上脚踏缫丝车了呢？

王老师：也是在宋代。人们在手摇缫丝车上加装了脚踏装置，宋朝人是世界上最早使用脚踏缫丝车的。

大　卫：这样的话，原先操纵摇杆的那只手就可以腾出来干别的了。

王老师：没错，别看只是解放了一只手，这可是缫丝工艺上的重大改革，大大提高了缫丝效率。

小　龙：脚踏缫丝车流行起来，是不是意味着可以提高丝线的产量？

王老师：对，明代开始普遍使用脚踏缫丝车，冷盆缫丝也成为当时缫丝的主流技术，两者的结合大大提高了缫丝效率。旁边这张图选自1871年出版的《蚕桑辑要》，上面画的就是脚踏缫丝车。

小　龙：脚踏缫丝车一直沿用到什么时候？

王老师：清代末期，也就是20世纪初。其实，现在还能看到这种缫丝车，不过都是小规模使用，主要是为了传承缫丝技艺。

大　卫：王老师，接下来就可以织锦了吧？

王老师：还早呢。宋锦的工艺复杂，光是准备工作就很多。要获得织锦用的丝线，下一步还得精练。

Treadle Reeling Machine

> Xiaolong and David visited Suzhou Silk Museum and met Mr. Wang, Prof. Long's friend, at the reception.

Xiaolong, David: Hello, Mr. Wang. Pleased to meet you.

Mr. Wang: Me too. Welcome to the museum! Prof. Long asked me to introduce some basic knowledge about Song brocade. So, what would you like to know?

Xiaolong: Thank you, Mr. Wang. Shall we start with the main process of weaving Song brocade?

Mr. Wang: Sure. Making Song brocade needs tons of work. A procedure of more than twenty steps would be carried out to make the end product. We can take a walk around while talking.

David: Okay. When we were visiting the sericulture and silk production centre I learned something about silk reeling. It is to extract silk fibres from cocoons, am I right?

Mr. Wang: That's true. The job of extracting and processing silk fibres is the first step of weaving brocades. Do you know how long a silk filament we can get from a cocoon?

David: It couldn't be very long, I guess, since a cocoon is so small.

Mr. Wang: Quite the opposite! Generally, the silk filaments from spring cocoons can reach a length of 900 to 1500 metres, while the filaments from summer and autumn cocoons can be around 700 to 1200 metres long.

David: Unbelievable! I didn't expect that.

Xiaolong: I remember that silk reeling was quite an effort. It involves boiling cocoons and seeking the end of silk fibre. It also needs special tools.

Mr. Wang: Both of you are quite knowledgeable indeed. Speaking of the tools used in silk reeling, the reeling machine is an important one worth mentioning.

Xiaolong: Reeling machine? I've never heard of it.

Mr. Wang: Look, here is a treadle reeling machine. It's a real-size model according to the description in ancient books.

David: I see. So, we have to use a foot to operate it, right?

Mr. Wang: Yes. It's actually an improved version after hundreds of years of development in silk reeling technology. In the very beginning, tools commonly used for silk reeling were basins and baskets. Back then, people would soak the silk cocoons in a hot water basin and then extract the silk threads by hand, and wind them onto the silk basket. The productivity was quite low. It was not until the Tang Dynasty that hand-reeling machines began to be widely used. With the machinery, workers brushed the unwound fibres in hot water and collected their ends while pushing the handle to spin the wheel, which helped draw the fibres out and wind them on the wheel.

Xiaolong: So, workers had their hands full. They had to reel the silk and boil the cocoons at the same time.

Mr. Wang: You're right, and they needed cooperation. As some workers reeled the silk, others would put the cocoons into the boiling water and control the fire by the side.

Xiaolong: Well, I believe they must take care of their pace then. If they put too many cocoons into the boiling water at a time and it would be impossible to reel them all in time, the cocoons would be over-boiled.

Mr. Wang: Exactly. You know, boiling cocoons will help soften the sericin that holds the cocoon together while preserving the fibroin, namely, the protein fibre of the silk. With the hand reeling machine, workers must do it quickly to prevent everything from dissolving away.

David: It seems to me that there's still room for improvement.

Mr. Wang: You're right. The ancients did come up with some ways of improvement. In the late Song Dynasty, a cooling basin was added to the silk reeling machinery in southern China. Workers would boil all the cocoons first and then put them into the cooling basin. After that, they could take their time to reel the silk.

David: What a clever idea!

Mr. Wang: It really is. In this way, the silk fibres taken out of the cocoon would be even in thickness, and gain more strength and lustre.

Xiaolong: What about other ways of improvement? Say, when did the treadle reeling machine appear?

Mr. Wang: Also in the Song Dynasty. Foot pedals were fitted to the hand reeling machine. That was the first invention of reeling device worked by feet in human history.

David: In this way, stepping on the pedals would set the machine in motion instead of moving the handle by hand.

Mr. Wang: That's true. The hand which used to push the handle could be otherwise occupied. It was really a significant technological renovation which greatly enhanced the production efficiency.

Xiaolong: With the popularity of the treadle reeling machine, the production of silk threads can also be increased, right?

Mr. Wang: Of course. Particularly in the Ming Dynasty, the combination of the treadle reeling machine and the cooling basin became common, with the efficiency of silk reeling greatly increased. Look, here is the treadle reeling machine illustrated in the book of *Key Facts about Silk* (《蚕桑辑要》) published in 1871.

Xiaolong: Is the treadle reeling machine still in wide use?

Mr. Wang: Well, it was kept in wide use up till the late Qing Dynasty or the early 20th century. Nowadays, occasionally we can still see this type of machine working for the purpose of preserving the ancient silk craftsmanship. But it's not in wide use anymore.

David: I see. Mr. Wang, what's next? Will it come to the weaving?

Mr. Wang: Not yet. As I mentioned earlier, weaving Song brocade is a complex process that involves a significant amount of work. Before that, we need to clean the silk fibres, or silk scouring, and prepare silk threads specifically for the weaving.

精练

小　龙：王老师，从蚕茧抽出来的丝为什么不能直接用来织锦呢？

王老师：抽出来的丝叫生丝，含有少量的蜡和灰分，有的上面还有油污。直接拿来织锦，会影响丝织品的质量。另外，生丝中含有大量的丝胶，不经过精练就用来织锦，会色泽暗淡，摸上去也缺少丝绸应有的顺滑感。

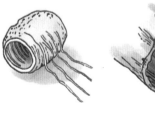
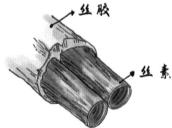

丝胶　Silk Sericin

大　卫：王老师，我不太明白。煮茧的时候，丝胶不是已经被热水软化溶解了么？

王老师：煮茧时丝胶的确会膨润，这样可以把丝从茧中抽离出来，但冷却后不少丝胶还会残留在生丝上。

大　卫：我明白了，精练是为了脱胶。

王老师：是的。不过织锦用的丝线不一样，需要精练的程度也不同。比如，宋锦中的面经用的是本色生丝，那就不需要脱胶，只需要去掉蜡质、油渍之类的杂质。而地经用的是熟丝，因为脱胶后的熟丝洁白柔顺，容易上色。

小　龙：看来给蚕丝脱胶是精练中很重要的工序。那具体怎么脱胶呢？

王老师：问得好。我们先说说灰练，这是古时常用的一种脱胶技术，就是用草木灰和蚌壳灰精练蚕丝。需要先把生丝放入草木灰、蚌壳灰液中浸泡，然后拧干水分，再涂上蚌壳灰，晚上把它们放入井水中静置。

大　卫：为什么要用草木灰和蚌壳灰呢？

王老师：草木灰和蚌壳灰属碱性，碱性水可以促进丝胶在水中溶解。

小　龙：古人还真是有办法。

精练 Silk Scouring 097

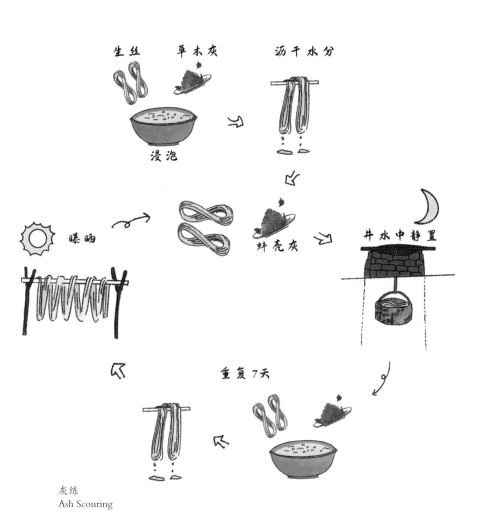

灰练
Ash Scouring

王老师：不过完整的流程可不是这么简单，刚才说的工序需要七天七夜连续重复，白天还得把沥过水分的生丝放在阳光下曝晒。

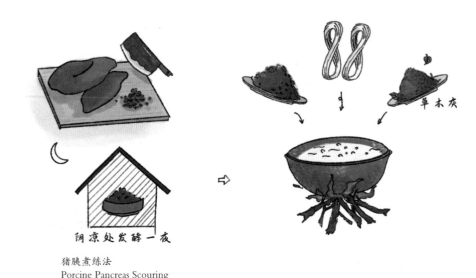

猪胰煮练法
Porcine Pancreas Scouring

大　卫：听上去还挺复杂的呢。您刚才说晚上要把丝线放进井水里，这又是为什么呢？

王老师：晚上把精练中的生丝放在井水中静置，主要是为了缓解白天曝晒以及浸泡碱水对丝素造成的损伤。

大　卫：原来是这样。那有什么不损伤丝素的好办法吗？

王老师：有。为了减少损伤，唐朝的时候人们在灰练基础上进一步改进，创造了猪胰煮练法。

小　龙：您是说要用猪的胰脏煮练丝线？

王老师：对。要把猪胰脏捣烂，放在阴凉处发酵一晚，然后和草木灰一起放进水中煮沸，再将生丝放进去精练。

小　龙：这是什么奇怪的原理呀？

王老师：一点儿都不奇怪，这是很科学的方法。猪胰煮练属于酶练。猪胰脏中含有一种胰蛋白酶，能水解丝胶蛋白。酶练比较温和，不会损伤丝素，脱胶效果也更好，还能增加蚕丝的光泽度。这可是个了不起的发明。中国是世界上最早使用酶练的国家。

小　龙：那现在的精练方法应该更多了吧？

王老师：是的。现代丝织业按照不同的需求，可以采取酸精练、碱精练，或者酶精练。这些不同的精练方法，可以大大提升精练的效率。

大　卫：王老师，生丝经过脱胶、除杂之后变为熟丝，下一步要干什么？

王老师：那就可以染色了。我们去织染坊看看吧。

Silk Scouring

Xiaolong: Mr. Wang, I have a question. Why don't we use the silk extracted from cocoons to weave brocade directly?

Mr. Wang: Here's the thing. The silk unwound from the cocoons is raw silk which carries impurities like wax and dust. Some even have oil stains, fats and dirt. If weavers use them directly, the quality of the product will be impaired. And raw silk contains sericin, which would reduce the lustre and smoothness of the brocade fabric.

David: But sericin should have been softened and dissolved in the boiling water in the previous step.

Mr. Wang: You're right. The heat of the boiling water can help to separate the silk from cocoons by hydrating water-soluble components in sericin. However, when the water cools down, some of the sericin may stay on the raw silk.

David: Now I understand that scouring is the step taken to remove impurities and sericin from the silk fibre, right?

Mr. Wang: Yes. The standard for silk scouring varies according to the kind of silk threads needed for the weaving. For example, the surface warp of Song brocade is made of natural raw silk, so the silk threads don't need degumming to remove the sericin. All it needs is removing impurities like wax and oils. The base warp, however, is made of silk that should be readily degummed and purified. This type of silk threads is white and soft, easy for dyeing.

Xiaolong: I see. It seems that degumming is an important step in silk production. Then, could you tell us how this process is carried out?

Mr. Wang: Sure. To start with, let's talk about a common degumming technique used in ancient China, that is, ash scouring. It involves using plant ash and clamshell ash to scour the silk. First, soak the raw silk threads in a solution of plant ash and clamshell ash. Then, gently drain off the water and coat them with clamshell ash. In the evening, hang the silk threads in a well.

Xiaolong: Why did they use plant ash and clamshell ash?

Mr. Wang: Well, in technical terms, they're alkalis that would facilitate the hydrolysis reactions of sericin.

Xiaolong: The ancients are really smart and resourceful.

Mr. Wang: Yes. But the complete procedure of ash scouring is time-consuming. The steps we just talked about should be repeated for seven days and nights, and after draining off the excess water from the silk, there should also be a long time of sun exposure during the daytime.

David: It sounds quite complicated to me. Mr. Wang, you also mentioned that at night people would hang the silk in a well. What's that for?

Mr. Wang: The well water can help to keep the silk fibres in a steady state at night and reduce the damage to the fibroin from sunburn and alkali.

David: Oh, that explains it. Besides ash scouring, are there any other ways that could do less damage to the fibroin?

Mr. Wang: Certainly. People in the Tang Dynasty made some improvements to ash scouring and introduced porcine pancreas scouring.

Xiaolong: You mean this time porcine pancreas was used in silk scouring?

Mr. Wang: Yes. Back then, people smashed porcine pancreas and left them fermenting in the shade for a night. After that, they put the fermented pancreas, plant ash and silk threads in boiling water for degumming.

Xiaolong: That's strange.

Mr. Wang: Not at all. It can be explained with scientific language. It's a type of enzymatic degumming. In the porcine pancreas, there's a substance called trypsin which helps dissolve sericin. Compared with other methods, enzymatic degumming is moderate and it would greatly lower the risk of damaging the fibroin. Also, it would not cause excessive degumming while securing the lustre of silk threads. China is the first country in the world to put enzymatic degumming into practice.

Xiaolong: What're the scouring methods available nowadays?

Mr. Wang: In modern silk weaving industry, various scouring methods such as acid scouring, alkali scouring, and enzymatic scouring can be used to meet different needs. The application of these different approaches has significantly enhanced the efficiency of the scouring process.

David: So, there's a great improvement as well. Now the silk has been degummed and purified. What comes next?

Mr. Wang: The process of dyeing. Let's go to the weaving and dyeing workshop.

茜草

> 小龙和大卫跟着王老师来到染织坊。

王老师：在丝绸织造中，染色是非常重要的一道工序。丝绸备受欢迎主要是因为它花纹别致，而且色彩漂亮。

小　龙：王老师，我有个问题。我知道世界范围内使用人工合成染色剂是19世纪之后的事。那以前丝是用什么染色的呢？

王老师：完全靠矿物和植物类的天然染料。人类使用天然染料的实践历史很长，考古发现在远古时期人类就开始使用这种染料画岩画了。

小　龙：我发现宋锦常常用同一色系的颜色，比如《彩织极乐世界图轴》里面就用到了各种红色，很难想象它们都是靠天然的染料染出来的。

> 王老师指着展柜里的茜草以及茜草不同生长期的图片，大卫认真看着说明。

王老师：人们在长期的实践中发现大自然里有着丰富的天然染料，也慢慢总结出各种染色技巧。我们先看看如何染红色吧。要染漂亮的红色可少不了这种植物——茜（qiàn）草。

大　卫：茜草，这个名字好听。

王老师：茜草是红色染料中的一种。"茜"字在汉语里是大红色的意思。

大　卫：可是，这说明上写着茜草是一种中草药。

王老师：茜草的确是一种能补血的中草药。不过在古代它还是一种重要的染料。在汉代，皇家服饰的主色调是茜红，所以对茜草需求量很大，他们会专门种植上千亩的茜草供染色使用。

小　龙：那规模真的很大。你们看，茜草的果实是红色的。染色用的应该是这种红果子吧？

王老师：那你就错了，用来染色的其实是它的根，不是果实。茜草根部的表面是黄红色，里面是紫红色，含有一种叫茜素红的化学成分。

茜草 Madder 107

茜草 Madder

大　卫：那得把里面的茜素红提取出来,对不对?

王老师：对。先要把茜草根切碎晒干,然后用热水把里面的茜素红煮出来。

大　卫：然后再拿来浸泡丝线就能染红色了,对吗?

王老师：没那么简单。一般来说,植物染料的着色性不稳定,直接用茜草根煮的水染色,只能染出浅黄色。

大　卫：那怎么才能染成红色呢?

王老师：得加入明矾。明矾入水后和茜素红发生化学反应,变成附着性很强的红色沉淀物。这样染成的红色丝线色泽鲜艳,织成锦后也不易褪色。

小　龙：那怎样才能染成深浅不同的红色呢?

王老师：按照传统色彩谱系的划分,茜草属于赤色系染料,用它可以染出从浅红、粉红到朱红、深红等一系列红色。古人通常是通过浸染次数控制颜色的深浅。比如,浸染四次可以染成朱红色。

小　龙：染色还真不简单呢。

大　卫：除了红色,宋锦中还有什么其他比较常见的颜色呢?

王老师：有啊,我们接着看看另一种常见的颜色——青色是怎么染成的。

Madder

Xiaolong and David followed Mr. Wang to the weaving and dyeing workshop.

Mr. Wang: Dyeing is a very important step in the weaving process. Chinese silk is popular not only for its exquisite patterns but also for its vivid colours.

Xiaolong: I have a question, Mr. Wang. The use of synthetic dyes worldwide began in the 19th century. But what was used to dye silk before that?

Mr. Wang: Well, in the past, the only dyes people could turn to were natural dyes extracted from minerals and plants. The use of natural dyes has a long history in human civilisation and some archaeological findings have revealed their use on cave paintings in ancient times.

Xiaolong: Really? But on Song brocade we often find different shades in use. For example, on the brocade scroll *Pure Land Brocade Scroll* we can see various shades of red. How is that possible to achieve solely with natural dyes?

> Mr. Wang pointed to the display cabinet containing pictures of the madder plant and its different stages of growth, and David read the explanations below.

Mr. Wang: Well, the natural sources for dyeing are quite rich. Over time, people gradually accumulated various dyeing techniques in practice. Now let's see how to produce a beautiful red colour. If you want to dye brocade red, madder is a necessity.

David: Madder, what a lovely name!

Mr. Wang: Madder is one of the most important natural red dyes. The character *qian* (茜) in its Chinese designation actually means a bright red colour.

David: I see. But it says here that madder is a Chinese herbal medicine.

Mr. Wang: It's true. Madder has a remarkable blood tonic effect. And it was also an indispensable dye in ancient China. For example, in the Han Dynasty, the dominant colour of imperial costumes was madder red. There was a great demand for madder back then. At that time, madders were planted in hundreds of acres for dyeing purposes.

Xiaolong: It was a large scale, I suppose. Look, the fruits of the madder are red. They are used for dyeing, right?

Mr. Wang: Not really. In fact, it's the root rather than the fruit that produces a red colour. Madder's roots are yellowish-red outside and red-violet inside. They contain a dye compound called alizarin.

David: So, it's necessary to extract alizarin from madder's roots, isn't it?

Mr. Wang: Exactly. The first step is to cut off the roots and have them dried. Then, alizarin is extracted by boiling them in hot water.

David: Now it's ready to be used to immerse and dye silk, right?

Mr. Wang: Not that simple. The pigmented substance produced in natural plant dyeing is found uneasy to adhere to the silk fibre. For example, dyeing with boiled water of madder's roots can only make silk threads light yellow.

David: Then how could they turn red?

Mr. Wang: We have to add alum. When alum dissolves in water, it reacts with alizarin to form red sediments which can dye silk with long-lasting red colours.

Xiaolong: I see. Then, how can we achieve different shades of red through dyeing?

Mr. Wang: According to the traditional colour spectrum in ancient China, madder belongs to a red dye that can produce a series of red colours ranging from light red and pink to vermilion and garnet. In ancient times, people obtained different shades of colour by adjusting the number of dyeing cycles. For example, if people intends vermilion, they would dye the fabric four times.

Xiaolong: I see. There's quite a lot to learn about the dyeing process.

David: Are there any other commonly used dyes in producing Song brocade, Mr. Wang?

Mr. Wang: Of course. Let's move on to indigo.

靛青

王老师：青色系是宋锦里很常见的颜色，也是中国传统色系里很重要的一种。

小　龙：可是我感觉绿色在丝绸中并不是常见的颜色呀？

靛青　Indigo

王老师：这里的青可不是绿色的意思。

小　龙：草色青青不就是草很绿的意思吗？

王老师：对，如果指草色，青的确是绿色的意思。但在染织物里，青指的是不同程度的蓝色。

大　卫：青色也是从植物中提炼的吧？

王老师：是的，青色来自一种叫靛青的染料，我们把制造靛青染料的植物统称为蓝草，有好几百个品种呢。2800年前，中国人就开始用蓝草染色了。

大　卫：这么早啊，也是用草根吗？

王老师：那倒不是。植物染色要根据植物自身的特点，选取不同的部分进行提炼，有的用叶和茎，有的用根，有的用果实，有的用花朵。用蓝草制作靛青，主要用叶和茎。

大　卫：哦，和红色的提取法不一样了。

王老师：对。制作靛青时，需要将新鲜蓝草的茎和叶浸泡在清水中发酵几天，等清水变成黄绿色后，再把茎叶滤走。

小　龙：然后也放明矾吗？

王老师：不是的。蓝色和红色的提取和使用方法不一样。染蓝色用的靛青是固态的。提取的方法是先将蓝草色素溶解,然后在绿水中加入生石灰粉搅拌,直到出现大量蓝紫色泡沫后进行静置,等水分蒸发完,沉在底部的固体就是染色要用的靛青。

大　卫:这个可需要耐心等待啊。可是固体的染料怎么用来染色呢?

五彩翟鸟纹宋锦
Song Brocade with Colourful Pheasant Design

靛青 Indigo 117

蓝地莲花纹宋锦　Song Brocade with Blue Lotus Pattern

王老师：需要将靛青和水按照一定的比例调和，然后再进行染色。比如，要染成色彩厚重的深蓝色，需要把丝线浸透染料后拧干、晾晒，让色素充分与空气接触氧化后再浸染。经过反复浸染后才能得到色泽均匀的深蓝色。

小　龙：原来深蓝色不是一次染成的，那就是说靛青可以染出不同程度的蓝色。

王老师：对。像宋锦中常见的天蓝和宝蓝，都属于传统的青色系，它们都可以用靛青染制。相比之下，宝蓝要更浓郁、鲜艳一些，是古代王公贵族服饰的常用色，和金色搭配起来显得特别高贵华丽。

小　龙：王老师，我想起来一句名言，叫"青出于蓝而胜于蓝"。它和从蓝草中提取靛青有关吗？

王老师：是的，这句话正是从当时的印染工艺中总结出来的。直到现在，人们还会经常用它比喻学生超过老师，后人胜过前人。

Indigo

Mr. Wang: Well, before we get down to this particular dye, let's talk about the traditional Chinese colour of *qing* (青) first. It's a common colour used in Song brocade and important in the traditional Chinese colour spectrum.

Xiaolong: But the colour *qing*, or green is not commonly seen in silk fabrics. Am I right?

Mr. Wang: Well, here, the Chinese character *qing* does not refer to the colour of green.

Xiaolong: Huh? Isn't the phrase *caose qingqing* (草色青青) in Chinese supposed to mean that grass is green?

Mr. Wang: Well, when we're talking about the colour of grass, *qing* does mean the colour of green. However, in the context of dyeing textiles, *qing* could refer to the varying shades of blue.

David: Oh. Is the colour of *qing* also derived from plants?

Mr. Wang: Yes. It comes from a dye called indigo. Plants that yield the indigo dye are collectively called blue grass in the literal sense. In China, the use of blue grass as a dye can be traced back to 2,800 years ago.

David: That's quite a long history! Is indigo also extracted from the roots of blue grass?

Mr. Wang: Not really. The parts where we obtain natural dyes from plants vary. It could be their leaves, stems, roots, fruits or flowers. As for blue grass, people mainly use its leaves and stems to make indigo dyes.

David: I see. It seems that the extraction methods of blue grass and madder are different.

Mr. Wang: Right. When producing the indigo dye, people have to soak the stems and leaves of fresh blue grass in water and let it ferment for a few days until the water turns yellowish green. After that, the stems and leaves will be filtered out.

Xiaolong: Is alum also used in the dyeing process?

Mr. Wang: Not this time. The indigo for blue dye is solid. To get the indigo, we need to add quicklime into the green solution, then stir it until a large amount of blue-purple foam pops up. Next, let it sit and wait for the water to evaporate so that sediment will be formed at the bottom. The sediment is the indigo for dyeing.

David: That requires a lot of patience. How is the solid indigo put into use for dyeing then?

Mr. Wang: We need to mix the solid indigo with water in certain proportion to create a liquid dye, and then it's ready for dyeing. For example, if we want to dye the silk threads dark blue, after the first dyeing, we should get the silk threads out to air after wringing out the excess water, and leave the pigment taking in oxygen. Then, repeat this process several times to ensure a deep and even dye absorption.

Xiaolong: I see. The colour of deep blue is not produced in a single dyeing. Otherwise put, the indigo dye can produce different shades of blue colour.

Mr. Wang: You're right. For instance, with indigo, we can dye the silk fabric sky blue or sapphire blue. Both of them are traditional Chinese colours. Compared with sky blue, sapphire blue is richer and brighter, and it makes a perfect match for gold. The combination was commonly seen on the costumes for the nobles in ancient China.

Xiaolong: Got it. And, a famous Chinese saying just came to my mind, Mr. Wang. It literally goes like blue grass yields indigo and also yields to it. Now I see the reason behind. It's about the tradition of natural dyeing.

Mr. Wang: Exactly! The saying was derived from the dyeing techniques of that time. Later, it means a student outperforming his teacher or a person who becomes better than his predecessors.

攀华

大　卫：王老师,丝线经过精练和染色后,下一步就可以织锦了吧?

王老师：那还没到呢,中间需要经过造机、设计纹样等一系列工序。今天没有时间讲这些专业技术,我们就一起看看宋锦专用的织机,参观一下织工的织锦过程

攀华
Panhua Process

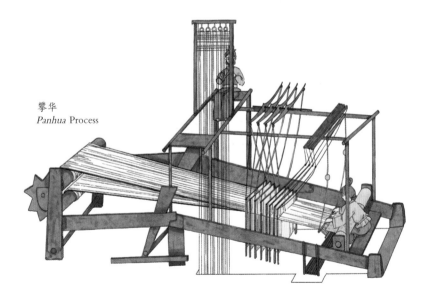

吧。织锦也叫攀华。

小　　龙：我记得在古汉语里，"华"就是"花"的意思，我猜在织锦里应该是指纹样或图案吧？

王老师：是这个意思。你们看，这就是过去手工织宋锦用的小花楼织机。

小　　龙：宋锦的织机在构造上是不是和其他类型的织机不一样呀？

王老师：过去织机一般是按功能设计，有的大，有的小，有的很复杂，有的比较简单。宋锦的织机是双经轴，这应该是它最大的特色。

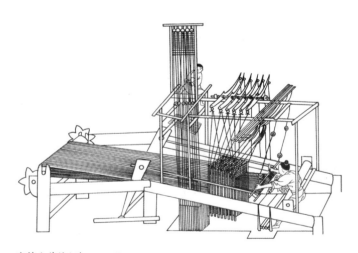

宋锦小花楼织机　Small Wooden Loom

小　　龙：这是不是和宋锦的经线、纬线联合显花，以及用双重经线有关？

王老师：没错，你懂的还真不少。为了联合显花，宋锦织造会用到地经和面经两组经线。这两组经线的材质、粗细以及组织结构都不同，各自的功能也不一样，因而宋锦织机采用上、下两个经轴，其中面经放在上轴，地经放在下轴。下轴带动地经织正面的纹样，上轴的面经帮助压住背面的纬线浮长。

大　　卫：为什么这么复杂？

王老师：这样可以维持张力的平衡，也就是说避免锦面因为线的粗细不一、结构不同而产生的不平整。宋锦织法是正面朝下，反面朝上。用双经轴织出来的正反面都很光滑，锦面平整。

大　　卫：明白了。呦，这个织机有上下两层呢，上面的人在做什么呢？他好像没有在织布呀。

王老师：你观察得很仔细。宋锦的组织结构相比以前的锦要复杂，还要呈现经线、纬线联合显花，所以织机结构也更复杂，织法更多样。得靠两个人在织机上相互配合，才能让经纬线完美交织，织出美丽的宋锦。

大　　卫：那具体怎么分工呢？

王老师：坐在下面的是织工，主要负责投梭打纬，也就是我们通常理解的织布。另外一个人是提花工，坐在织机中间悬空的花楼上，控制经线的起落。

大　卫：可是那么多经线，又该怎么控制呢？

王老师：这个要靠上面挂着的花本，就是那些绳结。织锦前要先设计好图案，并且按照图案上的纹样和颜色，用线编结成花本，挂在织机的上面一层，这些花本里面贮存着织锦的组织结构信息，让经纬线的排布有序。

小　龙：听上去很复杂呢，感觉像是给织机设定了一个编织程序。

王老师：对，你理解得完全正确。你们要是对花本感兴趣，可以去南京云锦博物馆看看，云锦里的妆花对这个技术运用得最好、最成熟。

大　卫：看着上面一团团线很普通，没想到里面还有大学问呢。

小　龙：真是大智慧呀。你们看，提花工一次次提起丝线，像不像鱼儿上钩的画面。

王老师：是的，这就叫"游鱼衔饵"。花本只是储存了经线起落的顺序，但没有提供花纹组织结构和色彩的信息。

大　卫：那色彩和花纹结构由下面的织工控制，对吗？

王老师：是的，小花楼织机上有一套束综和综片起落装置，综片的宽度根据织物的宽度而定。按照宋锦的组织结构和经线密度，一般采用6片综片分别控制地经与面经，通过综片的升降顺序进行打纬。

大　卫：这听上去也很复杂呀。

王老师：是的，非常复杂。我只是说了机器工作的原理，在你们看到他们这样织锦之前，还有好几道复杂的技术工序，需要多人合作完成。

大　卫：看来我们只是看到了完美技术的最后呈现。现在织宋锦还会用这些传统的工具吗？

王老师：如今，随着科技的进步，织造宋锦所用的提花机已经电子化了，效率高，花色多。

Panhua Process

David: After scouring and dyeing, silk threads can now be used for weaving, right?

Mr. Wang: Not quite yet. Before that, there're a series of steps involved such as machine preparation and pattern designing. We don't have enough time to go through all these particularities today. Let's start with the looms for weaving Song brocade and explore the *panhua*（攀华）process.

Xiaolong: I remember the Chinese character *hua*（华）originally means flowers. In the context of weaving brocade, it refers to patterns or designs. So *panhua* is a term referring to the weaving process, right?

Mr. Wang: It's true. Look at this. It's a traditional Chinese drawloom called Small Wooden Loom which was used for weaving Song brocade.

Xiaolong: Well, is it different from other looms in terms of structure?

Mr. Wang: Yes. Looms were generally designed based on their functions, varying in size and structure. For instance, the loom used for weaving Song brocade has a distinctive feature — double warp beams.

Xiaolong: Oh, does this have something to do with the combined patterning by the warp and weft threads, as well as the use of double warp threads in Song brocade?

Mr. Wang: Wow. You do know quite a lot. In order to achieve the effect of combined patterning, people use two groups of warp threads when weaving Song brocade. They are respectively the surface warp and the base warp. The two kinds are different in material, thickness and structure and play different roles in weaving. The Song brocade loom uses two warp beams, with the surface warp on the upper beam and the base warp on the lower beam. The lower beam controls the base warp to weave the pattern on the right side, while the upper beam helps overlay the weft which exceeds the needed length on the other side.

David: It's so complicated.

Mr. Wang: It really is. Since the two kinds of warp have different thicknesses and structures, a double warp axis can maintain the tension balance of different warp threads and avoid unevenness in the fabric. During the process of weaving, the right side of Song brocade is facing downward and the other side upward, ensuring that both sides of the brocade are smooth and even.

David: I see. Look, this loom is very high and has upper and lower parts. What is the person on the upper part doing? It seems that he's not weaving.

Mr. Wang: You are sharp-eyed. Compared with other types of brocade in history, Song brocade has a more complex structure. It involves the combination of warp and weft threads to weave delicate patterns. The loom structure is also more complex for the need to perform diverse weaving techniques. It requires two people to work together on the loom to ensure the perfect interlacement of warp and weft threads.

David: Got it. But how do they actually cooperate?

Mr. Wang: The person sitting below is primarily responsible for the process of weaving the fabric, throwing the shuttle and beating the weft. The other on the upper part of the loom controls the raising and lowering of the warp threads.

David: But there're so many warp threads. How do they work it out?

Mr. Wang: See? It all depends on the cords hanging above, or the pattern samples. Before the actual weaving process starts, a design draft needs to be drawn and then a pattern sample should be woven according to the arrangement on the draft. The pattern sample stores the structural information of the brocade, illustrating the sequence of warp and weft threads.

Xiaolong: Sounds quite complicated. It feels like programming a weaving process for the loom.

Mr. Wang: Exactly. If you are interested in pattern samples, you can visit Nanjing Yunjin Brocade Museum to see *zhuanghua* (妆花) brocade, the best example of this technique.

David: Wow, it's really impressive. I never expected there is so much knowledge behind those seemingly ordinary silk threads.

Xiaolong: The designing philosophy is truly brilliant. Look, the figure puller above keeps lifting the silk threads. Doesn't it resemble the scene of luring fish to bite the bait?

Mr. Wang: You're quite right. It's figuratively termed as "swimming fish taking the bait". But the pattern sample only stores the sequence of warp threads to be lifted. It does not provide detailed information about the structure and colour of the pattern.

David: Then who does this part? The weaver below?

Mr. Wang: Yes. On the drawloom, there are a set of accessories, one of them is known as reeds. The width of the reed depends on that of the fabric. Based on the structure and warp density of Song brocade, usually six reeds are used to control the base and surface warp threads, tightening the weft through their rise and fall.

David: It's really beyond my imagination.

Mr. Wang: Never mind. It takes time to figure it out. In fact, I've only explained the basics of looms. Before that, there are several more technical processes that require collaboration among individuals.

David: I was completely unaware of the intricate techniques behind. Are these traditional drawlooms still in use today?

Mr. Wang: Nowadays, with the advancement of modern technology, the drawlooms for Song brocade have evolved into electronic ones with higher efficiency and more possibilities of pattern varieties.

活色

王老师：说到花色多，在攀华过程中，工人们会经常用到一种关键工艺，叫作分段换色。

小　龙：我想起来了，我爸爸之前在给我讲细锦的时候提起过这个工艺。说通过分段换色可以在不增加锦面厚度的情况下，提升织物的色彩效果，这样织出来的

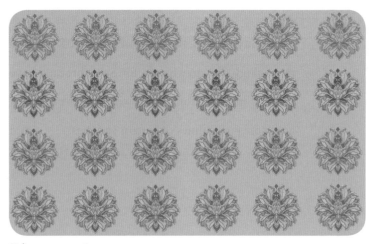

活色　*Huose* Technique

宋锦又薄又好看。

王老师：是的。这个工艺是在蜀锦基础上发展起来的。你们来看看这块宋锦样品。上面有很多重复出现的纹样，但是色彩却不同。这就是分段换色的效果。

小　龙：真好看，整体感觉灵动鲜活。

王老师：我们也把这个工艺称为"活色"，就是通过不断更换纬线的颜色变换花纹的色彩。

大　卫：这个名字很形象啊。颜色都活起来了！汉语太有意思了。

王老师：宋锦丰富的色彩，很大程度上得益于活色工艺。活色工艺的精妙，就在于既能表现丰富的色彩，又能保证织出的面料特别轻薄柔软。

大　卫：这是怎么做到的呢？

王老师：小龙，要不你先给我们讲讲？

小　龙：啊？其实我只是知道一点儿，那我先说说看。按照传统的织法，织锦的色彩是通过经纬线的一层层叠压显现出来的。为了增加颜色，需要叠加一层带有相应颜色的纬线。但这样一来，面料的厚度也随之增加了。宋锦织造工艺采取了不一样的思路，织造的时候在需要换色的位置不再叠加，而是换上新的纬线。至于说具体怎么织，我就不清楚了。

王老师：讲得不错。不增加叠压纬线还有一个优点，你们能猜到吗？

小　龙：是省线吗？这也算优点？

王老师：当然了，省了线，不就省了钱嘛。

大　卫：明白了，从经济角度上看，任何一种生产，降低成本最重要。

王老师：是的，这是宋锦技术的一个重要贡献。相比以前的锦，宋锦质地坚固，可以反复洗涤。

大　卫：这也算优点吗？

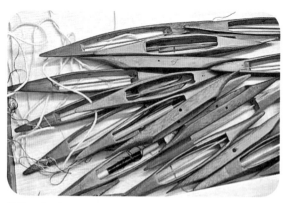

梭子　Shuttle

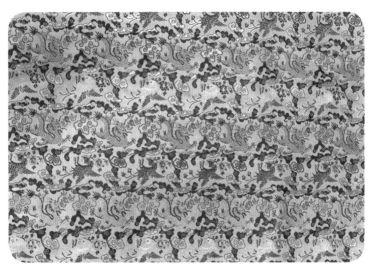

黄地龙凤云纹锦
Yellow-Ground Brocade with the Pattern of Chinese Dragon and Phoenix

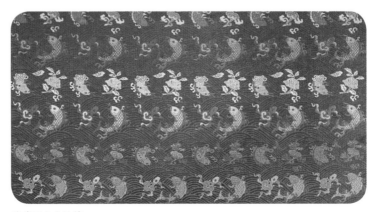

落花流水鱼纹锦
Fish Brocade with the Design of Floating Petals along Flowing Water

王老师：是的，很实用。说到织锦的具体方法，当然离不开一样重要的工具，那就是梭子。

大　卫：梭子我知道，就是织工手里拿的像个小船一样的东西。每个梭子上的彩色丝线就是纬线。梭子带着纬线在经线间往复，织出花纹。

王老师：没错。织宋锦的时候，用来分段换色的梭子可以更换不同颜色的纬线。需要换颜色时，就把织机停下来，在这一处换上另一种颜色的纬线。

大　卫：明白了，这样只在不同位置更换颜色就可以织出丰富的色彩了。

王老师：理解得不错！不过，织工偶尔也会在织物局部增加一重纬线来丰富色彩，但不会明显增加织锦的厚度。活色是传统织造技术的一大进步，后来被云锦织工采用，一直沿用至今。

小　龙：看来"活色生香"真是名不虚传。

王老师：是的，那接下来我们去看看当年的苏州织造署都做些什么工作。

Huose Technique

Mr. Wang: When it comes to the various colours of Song brocade, a crucial technique is often applied in the weaving process. It's known as colour variation in segments.

Xiaolong: Oh, I know that! When my father introduced fine brocade to me, he mentioned this term. He said this technique could enrich the colours of the fabric without adding extra weft threads and made Song brocade light and beautiful.

Mr. Wang: Yes. It was developed from the craftsmanship of Shu brocade of Sichuan province. Come here! Look at this piece. There are many repeated patterns, but they are different in colour. That's what this technique is about.

Xiaolong: It looks really beautiful and refreshing with rich colours.

Mr. Wang: This technique is also called *huose* (活色, literally lively colours coming to life) in Chinese, meaning changing the colours of the weft threads to create colour variations in the pattern.

David: Oh, this term is very vivid. It sounds like the colours are alive. The Chinese language is so interesting.

Mr. Wang: Yes. *Huose* technique plays an important role in making Song brocade more colourful. This technique is brilliant as it can add more colours to the fabric without adding weight or ruining the texture.

David: How is this possible?

Mr. Wang: Well, Xiaolong, why don't you tell us something about this technique first?

Xiaolong: Ah, I only know a little about that, but I'll give it a try. Using traditional weaving techniques, the colour of the brocade was created by interlacing warp and weft threads. To add one more colour, an additional set of weft threads needs to be added, rendering the fabric thicker. However, the craftsmanship of Song brocade features a different approach. Instead of incorporating new weft for the variation of colours, weavers replace existing threads with new ones. But I have no idea how that is done in detail.

Mr. Wang: Good job. Not adding new weft threads brings another advantage. Can you guess what it is?

Xiaolong: Saving threads? Is that an advantage?

Mr. Wang: Of course. Saving threads means saving money.

David: I see. From the economic perspective, reducing costs is crucial in any production process.

Mr. Wang: Exactly. This is an important contribution of Song brocade weaving technique. Besides, compared with other types of brocade in history, the fabric of Song brocade is durable and can be washed repeatedly.

David: Is that an advantage as well?

Mr. Wang: Certainly. When it comes to brocade weaving, we can't overlook an important tool—the shuttle.

David: I know it. It's a small boat-shaped object held by the weaver. The coloured threads on each shuttle are the weft threads. The shuttle carries the weft threads back and forth through the warp threads to produce different patterns.

Mr. Wang: You're right. Shuttles for segmented colour change can carry weft of different colours. Whenever it needs to change colour, weavers will stop the loom and start with a new weft colour.

David: I see. In this way, weavers only need to change weft colours in different sections, right?

Mr. Wang: Good. You got it. But weavers will occasionally overlay a set of weft threads in certain sections to bring some extra richness in its colour without adding too much thickness. *Huose* technique marked a major progress in traditional weaving techniques and was later incorporated into the craftsmanship of Nanjing Yunjin brocade. It's still an important weaving technique in use today.

Xiaolong: No wonder *huose* technique is seen as a prominent feature of Song brocade weaving. "Lively colours scent the pattern." This would be an apt description of Song brocade.

Mr. Wang: Absolutely. Now let's go and check out the Suzhou Weaving and Dyeing Bureau.

苏州织造署

> 小龙和大卫跟着王老师来到贡织院展厅。

大　　卫：这里和前面展厅风格完全不一样,好特别呀。

王老师：这个展厅是以清代苏州织造署遗址为原型设计的,介绍了清代官营丝织的历史,陈列的都是清朝宫廷和官员使用的丝织物。

大　　卫：原来是这样。苏州丝织的历史应该很长吧?

小　　龙：要是从宋朝算起,应该有1000多年了。

王老师：实际上要比那个时间长得多。2000多年前的春秋时期,苏州是吴国的都城,当时就是著名的丝绸产地。现在,古城区有一条叫锦帆路的街道,这个名字还和那时候的丝织有关系呢。传说当时吴王夫差[①]乘坐的船是用锦作风帆,后来河道变成了马路,人们就将此

苏州织造署
Suzhou Weaving and Dyeing Bureau

路命名为锦帆路。

大　卫：苏州丝织的历史竟然这么悠久。小龙，找个时间我们一起去锦帆路走走。

小　龙：好呀。这个吴王可真是奢侈，竟然用织锦做船帆。不过这也可以说明当时丝织业很发达，是吧？

王老师：是的，苏州在历史上一直是重要的丝绸产地之一。宋、元、明、清四朝都在苏州建有丝绸织造机构，督办各种名贵丝绸的生产,包括宋锦、龙袍等等。

大　卫: 原来是这样,那宋朝应该是宋锦最繁荣的时候吧?

王老师: 那倒不是。应该说,明清时期是宋锦的黄金时期。清朝初年,苏州织造署在生产规模上远超前朝,内部分工也更加细化,专业程度很高,是当时全国产量最高的织造中心。

大　卫: 当时还有其他的织造中心吗?

王老师: 有的,历史上南京、杭州都设立过织造署,三地的织造署并称"江南三织造"。

大　卫: 苏州织造署主要是为皇室织造丝绸吧?

王老师: 没错,这是它的主要任务。现在故宫收藏的十几万件织绣藏品中,苏州织造占了足足一半。其实织造署的职能不仅是为皇家提供丝织品,同时还要管理民间织造、征收机税。

小　龙: 您是说当时苏州城里还有很多老百姓也从事织造业?

王老师: 是的,当时由于皇室和贵族丝绸需求量大,官办织造没有那么大的生产能力,就得另外雇用民间机户。所谓民间机户,就是家里有织机的业主。

小　龙: 那就是官办带动民办了吧?

王老师：是的。这些机户有大有小，大的机户有几十台织机，小的机户就只有几台。他们生产的丝织品有的被官府收购，但大部分送到市场上去销售。

注释：
① 吴王夫差：？—公元前473，春秋时期吴国末代国君。

Suzhou Weaving and Dyeing Bureau

Xiaolong and David, following Mr. Wang, went to the exhibition hall of the Suzhou Weaving and Dyeing Bureau.

David: What a special exhibition hall! The style is completely different.

Mr. Wang: This hall is modelled on the relic of the Suzhou Weaving and Dyeing Bureau in the Qing Dynasty. It well demonstrates the history of the state-owned silk industry at that time. In the showcase, you can find many silk fabrics for the court officials of the Qing Dynasty.

David: Oh, that makes sense. Then Suzhou must have a long history of silk production.

Xiaolong: Yes. It began from the Song Dynasty, with a history of 1,000 years.

Mr. Wang: Actually, it goes back much further than that. During the Spring and Autumn Period over 2,000 years ago, Suzhou was the capital of the state of Wu and was already renowned for its silk production. Today, you'd still find a street known as Jinfan Road (锦帆路, literally Brocade Sail Road) in the old town. It's said that the sails of the boat for King Fuchai of Wu[1] were made of brocade. Later, the river course became a road, and hence the name Jinfan Road.

David: I never realised that silk weaving in Suzhou has such a long history. Xiaolong, we should take a walk on Jinfan Road later.

Xiaolong: I'd love to! It's quite extravagant of King Fuchai to use brocades as boat sails. But it actually speaks of the prosperity of the silk weaving industry then, right?

Mr. Wang: Indeed. Suzhou has always been an important place in the silk production history. During the Song, Yuan, Ming, and Qing dynasties, silk weaving institutions were established in Suzhou commissioning and managing various high-end silk production, including Song brocade and imperial robes.

David: I see. I guess the Song Dynasty probably witnessed the best years for Song brocade production in history, right?

Mr. Wang: Not exactly. Technically, the Qing Dynasty was the golden period for Song brocade. In the early years of the Qing Dynasty, the Suzhou Weaving and Dyeing Bureau developed on an unprecedentedly large scale. Back then, a more specific division of labour and a higher level of professionalism came into being and the Bureau became a weaving centre with the highest output across the country.

David: Were there any other weaving centres back then?

Mr. Wang: Yes. There were centres in the cities of Nanjing and Hangzhou. Together with the Suzhou Weaving and Dyeing Bureau, the three offices were known as the "Three Weaving Centres in the South of the Yangtze River".

David: So the Suzhou Weaving and Dyeing Bureau was mainly

responsible for weaving brocades for the imperial family, right?

Mr. Wang: You're right. That was its biggest job. Today, among the collections of the fabrics and embroideries in the Palace Museum, a full half of them were made in Suzhou. In fact, the Weaving and Dyeing Bureau not only provided silk products for the imperial family but also exercised regulation on privately-owned weaving workshops and levied taxes on them.

David: Do you mean there were many common people in Suzhou engaged in silk manufacturing?

Mr. Wang: That's true. Back then, the huge demand of the imperial family for silk exerted mounting pressure on the official weaving bureaus. Officials had no choice but to employ silk manufacturers among the local populace for greater production. These manufacturers employed were also owners of looms.

Xiaolong: In other words, the state-owned weaving bureau promoted and popularised the practice of silk fabric manufacturing, right?

Mr. Wang: You may say so. The manufacturing capacity varied. For example, big owners had dozens of looms whereas small owners only had a few. Some of the silk products they made were purchased by the government, but the majority were sent to the market for sale.

Note:

1. **King Fuchai of Wu:** ?—473 B.C., the last ruler of the State of Wu during the Spring and Autumn Period.

立桥　丝账房

王老师：清朝初年，苏州民间的织造业非常兴旺，一度出现了"东北半城，万户机声"的盛况。

大　卫：家家户户都在织锦，多么壮观的景象！为什么说是

立桥
Bridge Standby

东北半城呢?

王老师：当时在苏州聚集了大批分散的机户、机匠，他们大多居住在城东。

小　龙：机户、机匠听上去差不多嘛。

王老师：差很多呢。他们是雇佣关系，机匠受雇于机户。机户出设备，机匠出力气。在当时还有个有趣的现象，叫"立桥"。

大　卫：是立在桥头吗？谁立在桥头呀？

王老师：字面上确实是这个意思。"立桥"就是说成群的机匠每日清晨站在桥头附近，等待机户雇佣的场景。

大　卫：这个词还真形象。机匠们没有固定的雇主吗？

王老师：当时苏州城内的机匠主要分两类。一类是签了契约，有固定的雇主，按工作天数计算酬劳。另一类是零工，没有固定的雇主。

大　卫：那就是零工聚集在桥头了。可我还是不明白，为什么要聚集在桥头呢？

小　龙：这个我知道。苏州的河多，桥也多。现在部分老城区还保持着河街相邻的双棋盘格局呢。

王老师：是的。古时苏州是著名桥乡，桥是苏州的重要集会场所。清朝时期，晨聚晨散的桥市自然也就成了自由机匠的雇佣市场。

大　卫：原来是这样。我喜欢"立桥"这个说法，就像"活色"一样生动有趣。

王老师：的确是这样。不过，立桥也不是随便找个桥市去站着就行了，那样可找不到活干。机匠根据自己擅长的丝织种类，得站在不同的桥头等待雇主。苏州地方志里有这样的记载："花缎工聚花桥，素缎工聚白蚬桥，纱缎工聚广化寺桥，锦缎工聚金狮子桥。"

丝账房
Silk Trader

小　　龙：看来当时已经形成了招收不同工种的各种桥市，这应该也是清朝丝织业黄金时代的见证呀。

王老师：对。在苏州丝织全盛时期，民间丝织业相当发达，产销链也很成熟。这里除了机户和机匠，还有个重要的角色，叫"丝账房"。

大　　卫：丝账房？

小　　龙：那不就是算账发工钱的地方吗？

王老师：小龙望文生义了。我来说一下，你就明白什么是丝账房了。丝账房先购进原材料，如染好色的丝线，然后将这些原材料发放给机户织造。织成后，他们再收购丝绸制品，送到绸缎庄去售卖，从中获利。

小　　龙：我明白了。原来丝账房兼顾原材料供货商和经销商两个身份。他们没有织机，但有资金购买原材料，然后会委托机户生产，再把丝织品推销给卖家，丝账房作为中间商推动产销链的运作。

王老师：就是这个意思。其实，当时的丝账房在流通中扮演的角色不是那么绝对，也有一些丝账房会自行购进织机进行生产。

大　　卫：有了丝账房，加上机匠、机户、卖家，这样就齐了，可以形成一个比较成熟的商业经营体系。看来丝账房这个角色在当时的丝织业中必不可少呀。

王老师：这件事可不简单，这是中国本土现代工业萌芽的见证。

Bridge Standby and Silk Trader

Mr. Wang: In the early Qing Dynasty, the non-official silk weaving industry grew prosperous in Suzhou and in half the city in the northeast, the sound of looms could be heard from thousands of households.

David: Impressive! So many households were weaving brocades! But I am wondering why only the northeastern part was mentioned.

Mr. Wang: Well, that's because most loom owners and weaving workers lived in the east of Suzhou.

Xiaolong: What is the relationship between loom owners and weaving workers?

Mr. Wang: Weaving workers are employed by the loom owners. The owners provided looms and weavers made brocade fabrics. At that time, there was also an interesting activity known as *liqiao*（立桥）, or the bridge standby literally.

David: Bridge standby? Do you mean people would stand by a bridge? But why?

Mr. Wang: If you take it literally, yes. Specifically, it's about weavers standing around each end of the bridge every morning to get hired.

David: Oh, I see. What a vivid term! So weavers didn't have regular employers, right?

Mr. Wang: It depends. There were two types of weavers then in Suzhou. Contracted workers were regularly employed and paid for their working days, and those who had not signed a contract didn't work for a regular employer.

David: Now I get your point. Only freelance weavers gathered at the ends of the bridge, right? But why did they choose to gather around there?

Xiaolong: I know why. Suzhou has many rivers and bridges. Even today, some of the old towns in Suzhou maintain the double checkerboard patterns where waterways are intertwined with streets.

Mr. Wang: You are right, Xiaolong. In ancient China, Suzhou was known as "Home of Bridges". The bridge was once an important gathering place for the morning market with a high footfall. As a result, the ends of the bridge became an ideal location for freelance weavers to seek for employment.

David: Now I see. I am really interested in the Chinese term *liqiao* or bridge standby. It's such a picturesque description, just like the term, *huose* technique.

Mr. Wang: Yes. But standing by a bridge randomly would not help find a job. Weavers good at different silk weaving techniques would stand at different bridges to find the right job. As it was documented in Suzhou local chronicles, "Flowered satin weavers gathered at Flower Bridge, plain satin weavers at Baixian Bridge, damask weavers at Guanghua Temple Bridge and brocade weavers at Golden Lion Bridge."

Xiaolong: So, that means various bridge markets would recruit different types of weavers. This cultural scene was a proof of the prosperous silk industry in the Qing

Dynasty, I guess?

Mr. Wang: Absolutely. During this period, the privately run silk industry was already well developed. In addition to weavers and loom owners, there is another important role in this chain. They're known as silk traders.

David: What did they do specifically?

Xiaolong: Dealing with the trade?

Mr. Wang: Let me explain. A silk trader could be a person or an institution. Silk traders would first buy some raw materials like dyed silk threads and then distribute these materials to textile mills and loom owners. Once the weaving was finished, silk traders would then purchase the end products back and sell them to silk and satin shops to make profits.

Xiaolong: Got it. In this case, these silk traders are sort of today's dealers, right? They didn't have looms but they had money to buy raw materials, then outsourced the weaving work and finally acted as a supplier of the silk products. With this role in play, the production and marketing chain was activated.

Mr. Wang: Yes. Actually, silk traders at that time were quite versatile. Some would even buy looms themselves.

David: I see. Silk traders served as a connection between loom owners, weavers and sellers, and helped create an effective business model. It's quite important!

Mr. Wang: Indeed, its significance cannot be underestimated. The emergence of silk traders signalled the beginning of the early modern silk weaving industry in China.

雅文化

> 参观结束后,小龙、大卫和王老师来到博物馆的休息区。

王老师:小龙、大卫,我们今天的参观到此就结束了。休息一下吧。今天很匆忙,有些技术细节来不及展开讲,等下次你们再来博物馆,我们可以好好聊一下。

大　卫:谢谢王老师,我们下次再来向您学习宋锦织造的技术细节。

小　龙:王老师,我们今天收获可多了。我们弄清楚了宋锦的织造过程,从缫丝、精练和染色,再到活色技巧,感觉每一步都体现着古代匠人的智慧,太令人佩服了。

大　卫:我们还了解了不少古代织造业有趣的人和事呢,比

雅文化
Elegance Culture

如说立桥、丝账房。

王老师：有收获就好。相比其他织锦，宋锦的最大特色，莫过于它出众的雅致韵味。这一点除了得益于织造技艺的革新，还同宋朝士大夫阶层推崇的雅文化有关系。

大　卫：雅文化？

王老师：对，就是一种追求精致美好的文化现象。

小　龙：雅文化是相对于普通老百姓的市井文化而言吧？

王老师：是的。雅文化的背后体现的是有知识、有文化的文人士大夫群体的审美趣味和精神风貌。

大　卫：中国的琴、棋、书、画属于雅文化吧？

王老师：是的。古人把古琴音乐称为雅乐，吟诗、作画、下棋称为雅兴。"雅"字反映的是精神享受。在不同历史时期，雅文化的内涵有所不同。

大　卫：嗯，那宋朝的雅文化有什么特别的吗？

王老师：有啊。宋朝尚文治，雅文化非常盛行。要总结特点的话，那便是细腻雅致，含蓄内敛，自然和谐。

小　龙：我想起来了，宋朝有好几个皇帝都是书画爱好者，特别是北宋的宋徽宗[①]，是历史上有名的艺术家。他的花鸟画形神并举，书法独树一帜。是不是宋徽宗可以算作当时雅文化的代表?

王老师：对。宋徽宗不是个好皇帝，但的确是个优秀的艺术家。

小　龙：可以说宋代的雅文化对当时的宋锦织造产生了深刻影响，是吧？

王老师：没错。宋朝雅文化的审美情趣充分体现在丝织物图案设计上，艺术性很强。刚才你们在展厅里看到宋锦、蜀锦和云锦三大锦的专门展柜，有没有感觉宋锦的雅致风格很突出呀？

小　龙：是的，感觉蜀锦和云锦都很华丽，而宋锦是一种完全不一样的雅致风格。

王老师：宋锦色彩搭配沉稳、纹样风格柔美细腻，给人的感觉是平和有序、张弛有度，特别能体现中国文化的含蓄特质。

小　龙：老师，经您这么一解释，能感觉到宋锦有文人雅士的韵味了。难怪宋锦经常用作书画装帧的面料，真是再合适不过了。

大　　卫：谢谢王老师，带我们参观了宋锦织造，给我们讲解了宋锦相关的知识。

王老师：不客气，传承宋锦文化是我们的责任。欢迎你们以后再来参观。

注释：

① 宋徽宗：赵佶（1082—1135），宋朝第八位皇帝，擅长书法绘画，是古代著名书画家。

Elegance Culture

> After the visit, Xiaolong, David and Mr. Wang went to the rest area of the museum.

Mr. Wang: Well, take a rest, Xiaolong and David. Our visit today ended in a little bit of rush and we couldn't go through all the details. When you come next time, we can check out more details about Song brocade.

David: Definitely, Mr. Wang. We will come back next time to learn more about Song brocade from you. Many thanks indeed.

Xiaolong: We have learnt a lot today, Mr. Wang. We have now figured out how Song brocade is made. From silk reeling, scouring and dyeing to *huose* technique, every step embodies the wisdom of ancient Chinese craftsmen. It's truly admirable!

David: I agree. We also learnt some fun terms about the ancient silk weaving industry like "bridge standby" and "silk trader".

Mr. Wang: I am glad you find it rewarding. Compared with other brocades, the most prominent feature of Song brocade is its artistic flavour. It is derived not only from the innovation of weaving techniques but also from the elegance culture popular among the scholar-officials of the Song Dynasty.

David: What do you mean by elegance culture?

Mr. Wang: It's a phenomenon where people pursue an aestheticised way of life.

Xiaolong: I see. So, it's often seen as opposed to street culture, right?

Mr. Wang: Yes. It often reflects the aesthetic taste and spiritual demeanour of intellectuals and scholar-officials.

David: I see. Then the four arts of the Chinese scholar, *qin*(琴), *qi*(棋), *shu*(书) and *hua*(画) are also a part of this culture, I guess?

Mr. Wang: You're quite right. They refer to the arts of guqin, game of Go, Chinese calligraphy and painting respectively. In ancient China, people regarded *guqin* as an instrument of subtlety and refinement. And they took poem recitation, painting and playing Go as good taste. Elegance, or *ya* (雅) in Chinese reflects people's pursuit of spiritual enjoyment, and its connotation varies in different historical periods.

David: Well, is there anything special about the elegance culture in the Song Dynasty?

Mr. Wang: Of course. During the Song Dynasty, the ruling body valued cultural cultivation, and the elegance culture reached its peak during this time.

Xiaolong: Oh, I remember that several emperors in the Song Dynasty were big fans of calligraphy and paintings, especially Emperor Huizong[1], a famous artist in the Song Dynasty. He painted flowers and birds with a sense of unity both in form and spirit, and his calligraphy was renowned for its distinctive style. Can he be considered as an advocate of the elegance culture at that time?

Mr. Wang: Definitely. Perhaps he was not a competent ruler, but he was indeed an outstanding artist.

Xiaolong: In other words, the elegance culture in the Song Dynasty had a profound influence on Song brocade, right?

Mr. Wang: Yes. It influenced the creation of patterns and colour designs of Song brocade. We have just visited the special exhibition cabinets showcasing Song brocade, Shu brocade, and Yunjin brocade in the hall. Don't you feel that Song brocade really stands out?

Xiaolong: Right. Unlike the colourful and bright Shu brocade and Yunjin brocade, Song brocade has a completely different style of refinement.

Mr. Wang: The colours and patterns of Song brocade are designed in a reserved manner and its texture is exquisite, giving a sense of harmony, balance, and moderation. It particularly embodies the restrained beauty of Chinese culture.

Xiaolong: Thank you for your explanation. Now I can appreciate the elegance of Song brocade, Mr. Wang. No wonder it's often used as a material for binding books and framing paintings. What a perfect match!

David: Thank you for spending time with us, Mr. Wang. We have learnt so much about Song brocade.

Mr. Wang: You're welcome! It's my pleasure to share with you the culture of Song brocade. See you next time!

Note:

1. **Emperor Huizong:** 1082–1135, the eighth emperor of the Northern Song Dynasty who was a famous calligrapher and painter.

结束语

　　小龙和大卫参观了苏州丝绸博物馆，收获颇丰。他们不仅学习了宋锦织造工序，还深入了解了宋锦的历史文化。宋锦在继承蜀锦传统的基础上，形成了自身独有的特色。在设计上，宋锦兼具艺术审美与实用价值，在传统三大名锦中独树一帜。

Summary

Xiaolong and David had a fruitful visit to Suzhou Silk Museum. They not only gained a greater understanding of Song brocade weaving procedures and techniques, but also acquired some in-depth knowledge about its history and culture. The craftsmanship of Song brocade inherited the tradition of Shu brocade and developed its own distinctiveness. It boasts aesthetic and practical value in pattern design and colour scheme, standing out among the three famous types of traditional brocades in China.

中国历史纪年简表
A Brief Chronology of Chinese History

夏	Xia Dynasty			c. 2070—1600 B.C.
商	Shang Dynasty			1600—1046 B.C.
周	Zhou Dynasty	西周	Western Zhou Dynasty	1046—771 B.C.
		东周	Eastern Zhou Dynasty	770—256 B.C.
		春秋	Spring and Autumn Period	770—476 B.C.
		战国	Warring States Period	475—221 B.C.
秦	Qin Dynasty			221—206 B.C.
汉	Han Dynasty	西汉	Western Han Dynasty	206 B.C.—25
		东汉	Eastern Han Dynasty	25—220
三国	Three Kingdoms			220—280
西晋	Western Jin Dynasty			265—317
东晋	Eastern Jin Dynasty			317—420
南北朝	Northern and Southern Dynasties	南朝	Southern Dynasties	420—589
		北朝	Northern Dynasties	386—581
隋	Sui Dynasty			581—618
唐	Tang Dynasty			618—907
五代	Five Dynasties			907—960
宋	Song Dynasty			960—1279
辽	Liao Dynasty			907—1125
金	Jin Dynasty			1115—1234
元	Yuan Dynasty			1206—1368
明	Ming Dynasty			1368—1644
清	Qing Dynasty			1616—1911
中华民国	Republic of China			1912—1949
中华人民共和国	People's Republic of China			1949—

图书在版编目（CIP）数据

中国传统桑蚕丝织技艺 . 宋锦：汉英对照 / 刘润泽，董晓娜主编 . -- 南京：南京大学出版社，2024.8
（中国世界级非遗文化悦读系列 / 魏向清，刘润泽主编 . 寻语识遗）
　　ISBN 978-7-305-26376-7

Ⅰ . ①中… Ⅱ . ①刘… ②董… Ⅲ . ①桑蚕丝绸 – 锦 – 丝织工艺 – 介绍 – 中国 – 宋代 – 汉、英 Ⅳ . ① TS145.3

中国版本图书馆 CIP 数据核字（2022）第 234227 号

出版发行	南京大学出版社
社　　址	南京市汉口路 22 号　　邮编 210093

丛 书 名	中国世界级非遗文化悦读系列·寻语识遗
丛书主编	魏向清　刘润泽
书　　名	中国传统桑蚕丝织技艺 . 宋锦：汉英对照 ZHONGGUO CHUANTONG SANGCAN SIZHI JIYI. SONGJIN: HANYING DUIZHAO
主　　编	刘润泽　董晓娜
责任编辑	张淑文　　编辑热线　（025）83592401

照　　排	南京新华丰制版有限公司
印　　刷	南京凯德印刷有限公司
开　　本	880mm×1230mm　1/32 开　印张 6.125　字数 127 千
版　　次	2024 年 8 月第 1 版　2024 年 8 月第 1 次印刷
ISBN	978-7-305-26376-7
定　　价	69.00 元

网址：http://www.njupco.com
官方微博：http://weibo.com/njupco
官方微信号：njupress
销售咨询热线：（025）83594756

* 版权所有，侵权必究
* 凡购买南大版图书，如有印装质量问题，请与所购图书销售部门联系调换